The Maps of Canada

A Guide to Official Canadian Maps, Charts, Atlases and Gazetteers

Surveyors working on the Alaska-Canada Boundary in 1906. The surveyor on the left, H. F. Lambart, was a member of the first party to climb Mount Logan, Canada's highest peak.

The Maps of Canada

A Guide to Official Canadian Maps, Charts, Atlases and Gazetteers

N L NICHOLSON
Chairman of the Department of Geography, University of Western Ontario

L M SEBERT
Topographical Survey of Canada

First published in 1981

Wm Dawson & Sons Ltd, Cannon House
Folkestone, Kent, England

Archon Books, The Shoe String Press, Inc
995 Sherman Avenue
Hamden, Connecticut 06514
USA

British Library Cataloguing in Publication Data

Nicholson, Norman Leon
The Maps of Canada
1. Cartography—Canada
2. Canada—Maps
I. Title II. Sebert, L M III. Series
912′.71 GA473.7 79–41118

ISBN 0–7129–0911–7 (Dawson)

ISBN 0–208–01782–8 (Archon)

Printed and bound in Great Britain
by Mackays of Chatham

Contents

Figures

Tables

Preface

There is a reasonably abundant literature about Canada's map-makers whether it be in the form of biographies, the journals of the early explorers or accounts of the activities of the later surveyors and cartographers. But there has long been a need for a comprehensive survey of their products, i.e. the maps themselves. This we have attempted to produce.

It is abundantly clear that the web of mapping and charting of Canada is very complex but it is easy to justify for it is simply the result of a small nation in terms of population trying, frantically and desperately at times, to produce adequate maps of a federated state which covers half a continent. It is surely in this context that the present population density of 2.3 persons per square kilometre has meaning, and when mapping in the modern sense began officially the density of population was much lower.

Thus while the pages which follow are a guide through the maze of official map and chart series produced in Canada they also demonstrate the tenacity, flexibility, ingenuity and innovativeness of those responsible for the work.

In preparing this book we have had assistance from the research funds of the University of Western Ontario and the help and advice of many colleagues. We are especially grateful to Professor Bernard Gutsell the indefatigable editor-publisher of *The Canadian Cartographer* and Mr Serge Sauer, a past Chairman of the Geography and Map Division of the Special Libraries Association neither of whom we hasten to add, are responsible for any errors or omissions. We would also be remiss if we did not record our thanks to Diane Shillington and Irene Wayne who meticulously typed the manuscript (parts of it several times) and Derry Graves who drafted the illustrations.

London, Ontario — N. L. Nicholson

Ottawa, Ontario — L. M. Sebert

1. An Outline of the Mapping of Canada

The first maps of parts of what is now Canadian territory may well have been drawn long before Europeans first saw the coast of North America. The well developed cartographic sense of the Indians and Innuit of Canada is evident from the continuous record of mapping by these people since the beginning of European exploration. Most of their maps have survived only as copies made by Europeans, but there is no doubt that the cartographic intelligence of the Indians and Innuit was used by almost all European explorers who started the modern mapping of Canada with their often rough sketches.[1] Exploration in its very essence has its map-making element, for it is by the means of maps that the explorer describes where he has been and what he has observed.

In 1497 John Cabot made his famous landfall on what is now the east coast of Canada. As a navigator he had the necessary training to plot his ship's position from time to time, and the position and detail of the shore of the new land he was observing. From this data he drew at least one map but it has not survived. The only map which unambiguously incorporates John Cabot's discoveries is the world map by Juan de la Cosa dated 1500.[2] This is regarded as the earliest map which shows any part of Canada.

The same fate befell the maps of Jacques Cartier, which he himself drew as a result of his expeditions in 1534 and 1535, but his discoveries were promptly laid down by contemporary cartographers, some of whose maps have survived. The only known contemporary map showing the results of Cartier's first voyage without those of the second was drawn by Jean Rotz in 1535—the first map of the Gulf of St Lawrence region.[3] A map of circa 1537 incorporates the results of the first and second voyages. It is thus the first map of the Gulf and river St Lawrence, and the first on which the name Canada (and several other geographical names in use today) appears. It is not signed but is commonly attributed to Pierre Desceliers although it too may have been the work of Jean Rotz (Ganong, 1964).

The maps of the explorers further north fared no better. John Davis first penetrated the strait now named after him in 1585, but his charts have not survived. The first extant map record of his discoveries appears on the terrestrial globe of Emery Molyneux[4] dated 1592.

In any event, as the sixteenth century unfolded an increasing number of maps appeared in Europe showing with ever increasing detail the east coast of Canada.

The seventeenth century saw the arrival of Samuel de Champlain, an accomplished and zealous cartographer. Between 1604 and 1608 he produced no fewer than 11 large-scale charts of the approaches to harbours around parts of what is now the south-east coast of Canada, but he is perhaps best known for the four small-scale maps which were published between 1607 and 1632. Three of these were of New France and they map the Great Lakes for the first time. His maps were also drawn on skilfully constructed projections, ushering in the scientific era to Canadian cartography. Henry Hudson's discoveries of 1610–11 were incorporated into Hessel Gerritsz's 1612[5] chart of Hudson Bay and Strait and this was in turn incorporated by Champlain into his maps of New France. Champlain is the best example of the early inland travellers who were surprisingly good map-makers.[6] Others were not so talented. However, good or poor they all added to the growing knowledge of the vast land mass that stretched out before them as they progressed inland.

The early maps were generally drawn at small scales because the first travellers wanted to record the general lay-of-the-land in the regions through which they passed. In addition to setting out the geography of the country, such maps had a political or legal aspect in that they established a claim to sovereignty over the explored land.[7] This was of course important to the ruler who provided the expedition with financial support and who expected, eventually, some return on the investment in the form of furs, precious metals or other produce from the new-found territories. In Canada some explorers worked for a ruler more exalted than those of this world and looked for a spiritual harvest. The missionaries were often in the forefront of the penetration into new regions. In Canada government initiative and private enterprise have always worked hand-in-hand, and this was true in exploration. During the French régime the government explorers were closely followed, and at times overtaken, by the fur trading companies, and both were outdistanced by the private trader or *coureur de bois.* The fact that these latter gentlemen were in most cases trading illegally (the fur trade was supposedly a government monopoly) did not detract from their accomplishments in journeying through the land and bringing back information about distant places.

Many of the earliest maps were not noted for their geographical accuracy. The land was too difficult and the navigational training of the explorers was often too limited to produce gems of precision, but they did their best. Many of the explorers could calculate their latitude by measuring the altitude of the pole-star or the sun on the meridian, but longitude had to be estimated by the travel time in the westerly direction. On easy courses such as through the Great Lakes, reasonable accuracy was achieved. But over the rough terrain of the Canadian Shield, which lay as a barrier to the north and to the west, distance estimation was erratic, and as a consequence the longitude on the explorers maps was often seriously in error.

In the western interior of Canada a different pattern of exploration developed. It started in 1670 with the arrival on the shores of Hudson Bay of the Hudson's Bay Company with a charter giving it control of all lands draining into the Bay. For a time it restricted its operations to trade with the Indians who came down to the coast, but eventually it sent its employees inland to scout out the land. In 1691 Henry Kelsey became the first white man to see the Canadian Prairies. The French penetrated into the Prairies in the 1730s when the expeditions of La Vérendrye and his sons took them to the Saskatchewan River.[8]

However the journeys from their bases on Hudson Bay, in the case of the British, and Montreal, in the case of the French, were so arduous that in the early days few made the journey. The Prairies were so vast that there was no conflict of interest or of arms in the west before the end of the French régime in 1763. The arrival of the British in Quebec saw the start of detailed mapping and charting by such eminent cartographers as Joseph Des Barres and Samuel Holland. The work of the first culminated in the publication in 1777 of a magnificent collection of charts and plans of the coast of Nova Scotia and Cape Breton Island. Samuel Holland who became Surveyor-General of the British Colonies north of Virginia will be remembered as 'the author and finisher of the first organized scientific land and water surveys on the northern half of North America'.[9] Control of New France also gave British trading companies a new route into the Canadian west, and in the 1770s the Hudson's Bay Company, for the first time, was confronted by serious competitors. The governors of the Company realized that they had control over a vast land with uncertain boundaries and, to a large extent, unknown topography. The competition from traders using the southern route into their territory made them realize that they would either have to set up an inland trading network or let the cream of the fur crop filter away through the Great Lakes route to Montreal. The inland network was of course the only answer. In 1771 Samuel Hearne travelled down the Coppermine River to its mouth on the Arctic Ocean and produced a map of the area north-west of Churchill that was to remain for more than a century the only important source of information on this part of the country.[10]

But if the gentlemen in London were to make correct decisions regarding the location of supply routes and trading posts in central Canada, they needed more accurate maps. 'Inland surveyors' were hired. These were men specially trained at navigation schools in England to take observations on the sun and stars for both latitude and longitude, and by using these points as a framework they would draw route maps of commendable accuracy. Philip Turnor was one of these and in 1778–79 he produced the first map of the western interior of Canada based on scientific instrumental surveys.[11] Sir Alexander Mackenzie left an accurate cartographical record of his routes down the Mackenzie River in 1789 and across the Rocky Mountains to the Pacific Ocean in 1793.[12] David Thompson, one of the most famous and accomplished of Turnor's pupils, himself mapped with particular detail and accuracy, a remarkably large part of the west, much of it in mountainous country.

The west coast of Canada, like the east coast, was first explored from the sea. The extension of British sea power into the Pacific resulted in the detailed mapping of this region. James Cook had already conducted hydrographic surveys of the island of Newfoundland, Labrador and the St Lawrence River before his last voyage (1776–79) took him to the northwest Pacific. His mapping of this region was continued by one of his officers, George Vancouver, who in 1793 arrived off the island that now bears his name and commenced the detailed charting of the area for the British Admiralty.

A composite picture of the accomplishments of the early exploration of Canada is furnished by the remarkable map produced by the London map-maker John Arrowsmith in 1802.[13] On this map, Canada as a whole is beginning to take shape. Admittedly there are gaps in the knowledge; northern Quebec is a blank and the northern mainland coast from Hudson Bay to Alaska is entirely missing except for the two isolated points which Samuel Hearne and Alexander Mackenzie had reached. The extent and 'grain' of the Rocky Mountains is correctly depicted, and the Pacific coastline, thanks to the surveys of the British Admiralty, is shown in admirable detail. Arrowsmith and other European map-makers followed the unfolding of Canada with a series of new editions which were produced whenever the news of important new discoveries reached England. The exploration of northern Canada continued with men such as Edward Parry (1819–25) and Otto Sverdrup (1898–1902) discovering and mapping significant areas of new land before the end of the century.[14]

But even as the country just beyond the Great Lakes was being penetrated for the first time the second phase of mapping was underway in the east. This was the more detailed surveying and mapping of the areas where settlers and their families were establishing themselves in the small coastal colonies being set out by the colonizing governments. The allocation and administration of the colonial land, both French and British, required both a survey system and a land registration system. The surveyors marked out the land, and from their field notes produced the first plans of townships and seigneuries. Eventually these plans, which were kept in registry offices and in the files of government surveyors, were assembled into county and district maps. These are described in Chapter 2. In 1842 the Geological Survey of Canada was founded, and as geologists must have maps on which to display the results of their field investigations this small agency had from its inception a mapping capability. Originally its maps depicted the topography only as a base under its geological maps, but eventually a need developed for the topographic maps themselves. The early topographic maps of the Geological Survey of Canada are described in Chapter 4.

In 1867 the Confederation of Canada came into being, and shortly after (1869) it took control of the vast Hudson's Bay Company's lands. Settlement of this region was of prime concern because the American move west was in full flow south of the border, and the officials in the Canadian government could plainly see that this valuable farm and ranch land could not be held if it remained unoccupied. So in 1871 the Dominion Land Survey was begun, as described in Chapter 3. The Dominion land surveyors worked out of the Department of the Interior in Ottawa. Originally the survey agency within this department was called the Surveys Branch, but in 1883 the name was changed to the Technical Branch. In 1890, in recognition of the fact that the surveyors of the Branch were doing topographical surveys in the Rocky Mountains as well as cadastral surveys on the plains, the name was changed once more to the Topographical Surveys Branch.[15]

An examination of the maps produced in Canada up to the year 1900 provides evidence of great cartographic activity in all parts of the Dominion. But very little of this activity was directed to the production of detailed topographic maps. The federal government began to seriously assume the responsibility for hydrographic surveying in 1883 with the establishment of the Georgian Bay Survey and the publication of its first chart in 1886. Its development is described in Chapter 12. The rapid settlement of western Canada and the active encouragement of immigration saw the need for and the appearance of thematic maps and atlases (see Chapters 11 and 14). The cadastral work continued to support the homestead movement on the Prairies, and to a lesser extent in the eastern provinces. The Geological Survey of Canada worked throughout the country producing maps to support geological reports. Explorers ranged through the northland and gathered data which filled in the blanks of the small-scale maps of Canada. The three-mile maps on the Prairies covered that region in an orderly fashion with sheets of standard size and regular sheet lines. In an effort to distil the geography of the rest of Canada into a similar series, what became known as the Chief Geographer's series, at the surprisingly metric scales of 1:250,000 and 1:500,000, was commenced in 1903 (see Chapter 10).

In the same year the General Staff of the Canadian militia decided to take a close look at the state of topographic mapping in Canada. A British expert in this field, Major E. H. Hills, was invited to come to Canada and examine the situation at first hand. After a detailed study of existing maps, and the plans for mapping in the immediate future, he recommended that the military should start a methodical topographic mapping of the whole country.[16] His recommendation was accepted, and in 1904 the Survey Division of the Department of Militia and Defence was formed. The work of this division which was mainly at the scale of 1 inch to 1 mile is recounted in the chapters dealing with topographical mapping. In 1908 the Geological Survey of Canada decided to give recognition to the topographic surveying being done by its members by establishing the Topographical Division within the agency. So within the federal government structure from 1908 onward, there were three distinct units concerned with topographical mapping:

1 The Topographical Division, Geological Survey of Canada
2 The Topographical Surveys Branch, Department of the Interior
3 The Survey Division, Department of Militia and Defence

These agencies operated independently and there was little coordination in map design, or in areas of responsibility.

In 1922 the Board on Topographic Surveys and Maps was formed to overcome this and to bring about some standardization in Canadian topographic mapping. The design and sheet layout of the Militia one-inch maps were adopted by the Board which then went on to work out the different scales of the National Topographic System.

The two civilian topographic agencies were combined in 1936 into the Topographical Survey of Canada within the Bureau of Geology and Topography, which was itself in the Department of Mines and Resources (the successor to the Department of the Interior). The name of this department has undergone two further changes, to Mines and Technical Surveys in 1949 and to Energy Mines and Resources in 1966, but the Topographical Survey of Canada remains unchanged in title to this day. Its methods of mapping have of course changed dramatically, particularly with the advent of aerial photography as will be recounted in the chapters which follow.

Canada's military mapping agency has remained independent of its civilian counterpart. It was known as the Geographical Section of the General Staff from 1924 until 1946 when it became the Army Survey Establishment. In 1966 the Canadian government unified the Army, Navy and Air Force into a single service, and at that time the new title of the Mapping and Charting Establishment was adopted by the military mapping agency.

In the meantime the map of Canada was being completed. 'New land' had been discovered in the north by Vilhjalmur Stefansson in 1915–16, and as recently as 1948 additional land was being added to Canadian maps when, as a result of aerial surveys, three large new islands with a total area of 6000 square miles (15,000 sq km) were shown to exist.[17] Thematic mapping has increased enormously in recent years particularly since the appearance of the third edition of the *Atlas of Canada* in 1958. Such mapping is now carried out in a number of federal government departments. When their output is added to the mapping programme of each of the provincial governments, as is described in Chapter 13, the need for a guide to Canadian mapping is apparent. The chapters which follow are intended to fill this need.

2. Systematic Mapping before 1890

This chapter describes three styles of mapping that were precursors of the large map series that eventually would cover great areas of Canada. The year 1890 may be looked on as a watershed because it was during that year that the three-mile series, the first of the large series, was initiated (as will be outlined in the following chapter).

The British colonies in post-revolutionary North America needed maps. The very size of the country, the great distances between settlements, the difficulties of travel, all pointed to this fact. There was, however, no money in the colonial budget for general-purpose mapping. Any mapping that was done either had to have an immediate and clearly defined use, or had to be produced without the assistance of public funds. The first maps to be described were produced by virtue of the second of these options. These are the county maps of eastern Canada which were published by small map companies or, in a few cases, by private individuals. They may be considered as 'official' maps both because they were based on official surveys and because they provided the only detailed mapping available for many years. The military maps covered in the second section were needed to prepare for the defence of British North America against possible aggression by a newly-emerged military colossus, the Union Army. The final section of this chapter is devoted to the first detailed mapping of the Canadian Rocky Mountains. This small but interesting series came into being as a cartographic experiment when it became obvious that the survey methods that had worked so well on the Prairies could not be used in the mountains. The ingenious solution used to draw these maps resulted in elegant little sheets that were sufficiently accurate for the needs of the day, and yet were economical in terms of expenses for field surveys.

County Maps and County Atlases

In both the British and French systems of colonization surveyors were employed to lay out farm lots for the settlers who were to occupy the farmland of the colony. An integral part of

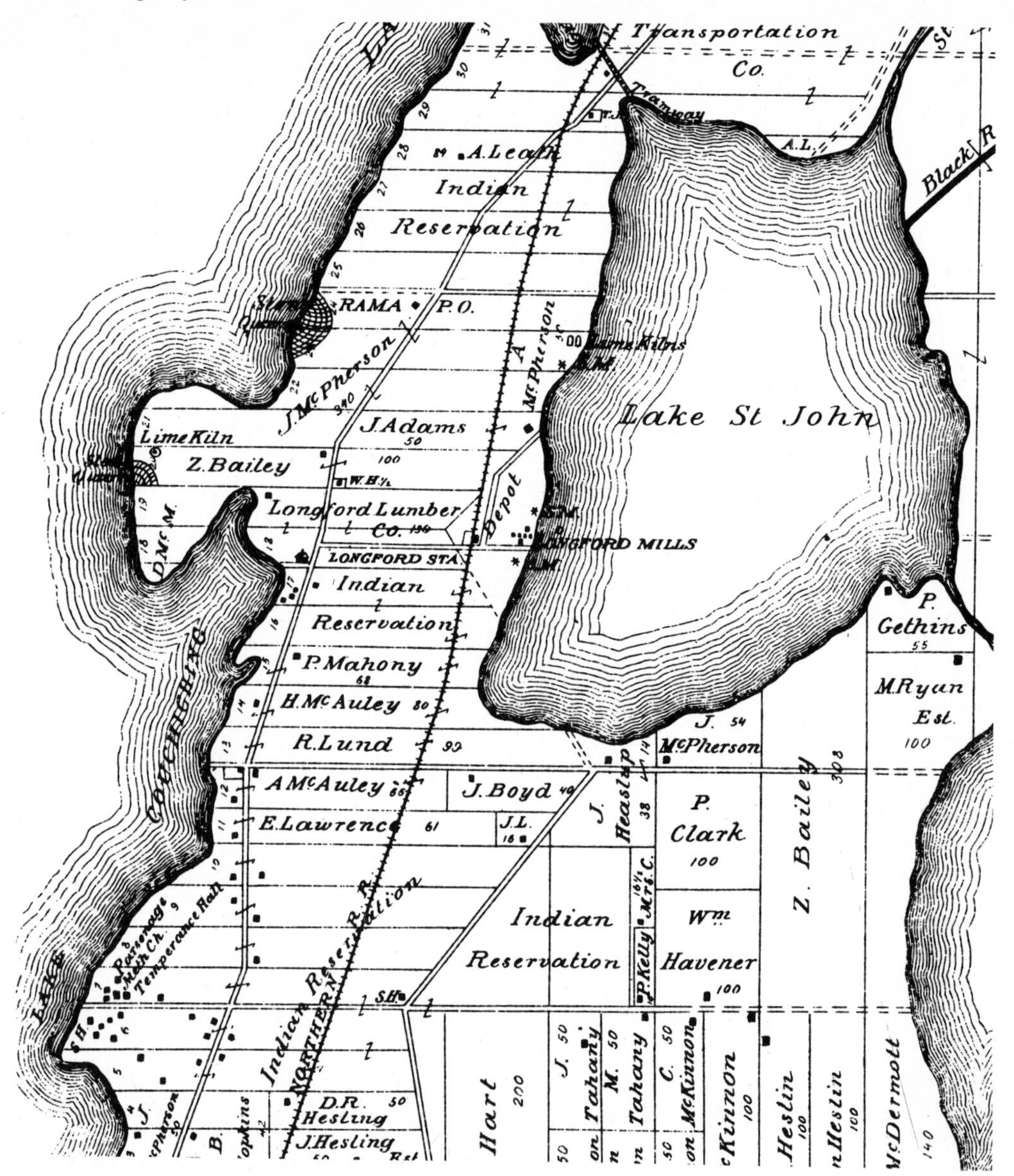

Figure 1 Section from Atlas of County of Ontario, 1875

the survey was the production of a plan showing the boundaries of the farm lots, the compass directions of the boundaries, and an outline of the more important topographical features so that the boundaries could be related to the ground. These plans were normally kept in the land administration headquarters of the colony where they were used as part of the record of land use and land ownership. The unit of land administration in French Canada was the seigneury and in the British colonies the township. The original township and seigneury plans were kept in the custody of the Surveyor-General of the province, but copies were made for a variety of purposes. In Upper Canada when the province was divided into counties and districts a certain amount of autonomy was given to the counties in the matter of land granting and administration. County surveyors were appointed and they were provided with copies of the township plans of their county. As new townships were opened, they in turn provided the Surveyor-General's office with hand-drawn copies of the new plans and at both county and provincial level county plans were compiled from the township plans. Certain officials of government and people of influence within the province could obtain copies of the plans of their county, but they were not on sale to the public.

The first detailed maps made available to the public were printed maps of the more heavily settled counties of Ontario. These were produced by private enterprise but with the tacit assistance of the provincial Surveyor-General who made the official plans (or copies thereof) available to the individual or company that was going to undertake the map production. One of the earliest of these maps was that of Prince Edward County published by Publius V. Elmore, a local surveyor, in 1835. It was engraved by the New York firm of Stiles and Co. at 90 chains to the inch, and shows the township survey patterns, the roads and trails that had been opened by the settlers, and the commercial and social facilities such as stores, mills, churches, wharves, etc. This type of mapping was much more advanced in the United States where companies were formed to put county mapping on a more substantial business footing. With the formation of these companies came a more serious marketing approach. The maps were sold by subscription before they were printed, and the subscribers were given recognition on the map by the inclusion of their names in prominent type on their farm lot or beside their place of business. For an additional fee drawings of the subscriber's farmhouse, factory or store would be included in the margin.[1] The county maps were wall maps varying in size from 100 to 150 cm in both dimensions depending on the size and shape of the county. Scales were generally given in chains to the inch, and these varied from 1 inch to 40 chains (1:31,680) to 125 chains (1:99,000). The scales of the more important county maps are given in Table 1.

Table 1 County Maps

Province and County	Printing date	Publisher	Scale
New Brunswick			
Carleton	1876	Roe and Colby	1 inch to 400 rods (1:79,200)
St John and King's	1862	W. E. & A. A. Baker	1 inch to 400 rods (1:79,200)
Westmorland and Albert	1862	W. E. & A. A. Baker	1 inch to 400 rods (1:79,200)
Northumberland	1876	Roe and Colby	1 inch to 500 rods (1:99,000)
Nova Scotia			
Annapolis	1876	A. F. Church	1 inch to 80 chains (1:63,360)
Antigonish	1878	A. F. Church	1 inch to 80 chains (1:63,360)
Cape Breton	1874	A. F. Church	1 inch to 80 chains (1:63,360)
Colchester	1874	A. F. Church	1 inch to 110 chains (1:87,120)
Cumberland	1873	A. F. Church	1 inch to 120 chains (1:94,040)
Digby	1871	A. F. Church	1 inch to 80 chains (1:63,360)
Guysborough	1876	A. F. Church	1 inch to 128 chains (1:101,376)
Halifax	1865	A. F. Church	1 inch to 500 rods (1:99,000)
Hants	1871	A. F. Church	1 inch to 80 chains (1:63,360)
Inverness	1883	A. F. Church	1 inch to 80 chains (1:63,360)
King's	1864	A. F. Church	1 inch to 80 chains (1:63,360)
Lunenburg	1883	A. F. Church	1 inch to 80 chains (1:63,360)
Pictou	1867	A. F. Church	1 inch to 80 chains (1:63,360)
Queen's	1888	A. F. Church	1 inch to 80 chains (1:63,360)
Richmond	1887	A. F. Church	1 inch to 80 chains (1:63,360)

Table 1—*continued*

Province and County	Printing date	Publisher	Scale
Shelbourne	1882	A. F. Church	1 inch to 80 chains (1:63,360)
Victoria	1887	A. F. Church	1 inch to 80 chains (1:63,360)
Yarmouth	1871	A. F. Church	1 inch to 80 chains (1:63,360)
Ontario			
Brant	1859	George C. Tremaine	1 inch to 40 chains (1:31,680)
	1859	George C. Tremaine	1 inch to 40 chains (1:31,680)
	1875	T. S. Shenston	1 inch to 50 chains (1:39,600)
Carleton	1863	D. P. Putnam	1 inch to 1 mile (1:63,360)
Durham	1861	George C. Tremaine	1 inch to 50 chains (1:39,600)
Elgin	1864	George R. Tremaine	1 inch to 60 chains (1:47,520)
Essex	1877	R. M. Tackaberry	1 inch to 180 rods (1:35,640)
Frontenac, Lennox and Addington	1860	Putnam & Walling	1 inch to 400 rods (1:79,200)
Haldimand	1863	W. Jones	1 inch to 50 chains (1:39,600)
Halton	1858	George C. Tremaine	1 inch to 40 chains (1:31,680)
Huron	1862	R. W. Hermon, R. Martin and L. Bolton	1 inch to 80 chains (1:63,360)
Kent	1876	Shackleton & McIntosh	1 inch to 60 chains (1:47,520)
Lanark and Renfrew	1863	D. P. Putnam	1 inch to 1.5 miles (1:95,040)
Leeds and Grenville	1861	Putnam & Walling	1 inch to 1 mile (1:63,360)
Lincoln and Welland	1862	G. R. & G. M. Tremaine	1 inch to 50 chains (1:39,600)
Middlesex	1862	G. R. & G. M. Tremaine	1 inch to 60 chains (1:47,520)
Norfolk	1856	George C. Tremaine	1 inch to 60 chains (1:47,520)
Ontario	1860	George C. Tremaine	1 inch to 50 chains (1:39,600)
Oxford	1857	George C. Tremaine	1 inch to 60 chains (1:47,520)
	1896	L. Bolton & Son	1 inch to 1 mile (1:63,360)
Peel	1859	G. R. & G. M. Tremaine	1 inch to 50 chains (1:39,600)
Perth	1888	L. Bolton	1 inch to 1 mile (1:63,360)
Prince Edward	1835	P. V. Elmore	1 inch to 90 chains (1:71,280)
	1863	George C. Tremaine	1 inch to 50 chains (1:39,600)
Simcoe	1871	John Hogg	1 inch to 80 chains (1:63,360)
Stormont, Dundas Glengarry, Prescott and Russell	1862	D. P. Putnam	1 inch to 80 chains (1:63,360)
Victoria	1877	Tom Kains	1 inch to 50 chains (1:39,600)
Waterloo	1861	G. R. & G. M. Tremaine	1 inch to 50 chains (1:39,600)
Wellington	1861	Leslie and Wheelock	1 inch to 60 chains (1:47,520)
	1885	Wm. W. Evans	1 inch to 60 chains (1:47,520)
Wentworth	1859	Hardy Gregory	1 inch to 50 chains (1:39,600)
York	1860	George C. Tremaine	1 inch to 60 chains (1:47,520)
Prince Edward Island			
Map of whole island	1863	W. C. & H. H. Baker	1 inch to 50 rods (1:99,000)
Quebec			
Deux-Montagnes	1888	J. H. Leclair	1 inch to 15 arpents (1:34,530)
St Francis District	1863	Putnam & Gray	1 inch to 400 rods (1:79,200)
Shefford, Iberville Brome, Missisquoi and Rouville	1864	H. F. Walling	1 inch to 300 rods (1:59,400)
Terrebonne	1886	J. H. Leclair	1 inch to 30 arpents (1:69,060)

County atlases were a direct consequence of the county map production. According to legend, one of the American mapping companies found that some of the purchasers of their maps had cut up the map, township by township, and pasted the pieces into an album together with hand-written notes on the size, population and business activity of each township. The sales advantages of having a product that would fit into the drawing-room

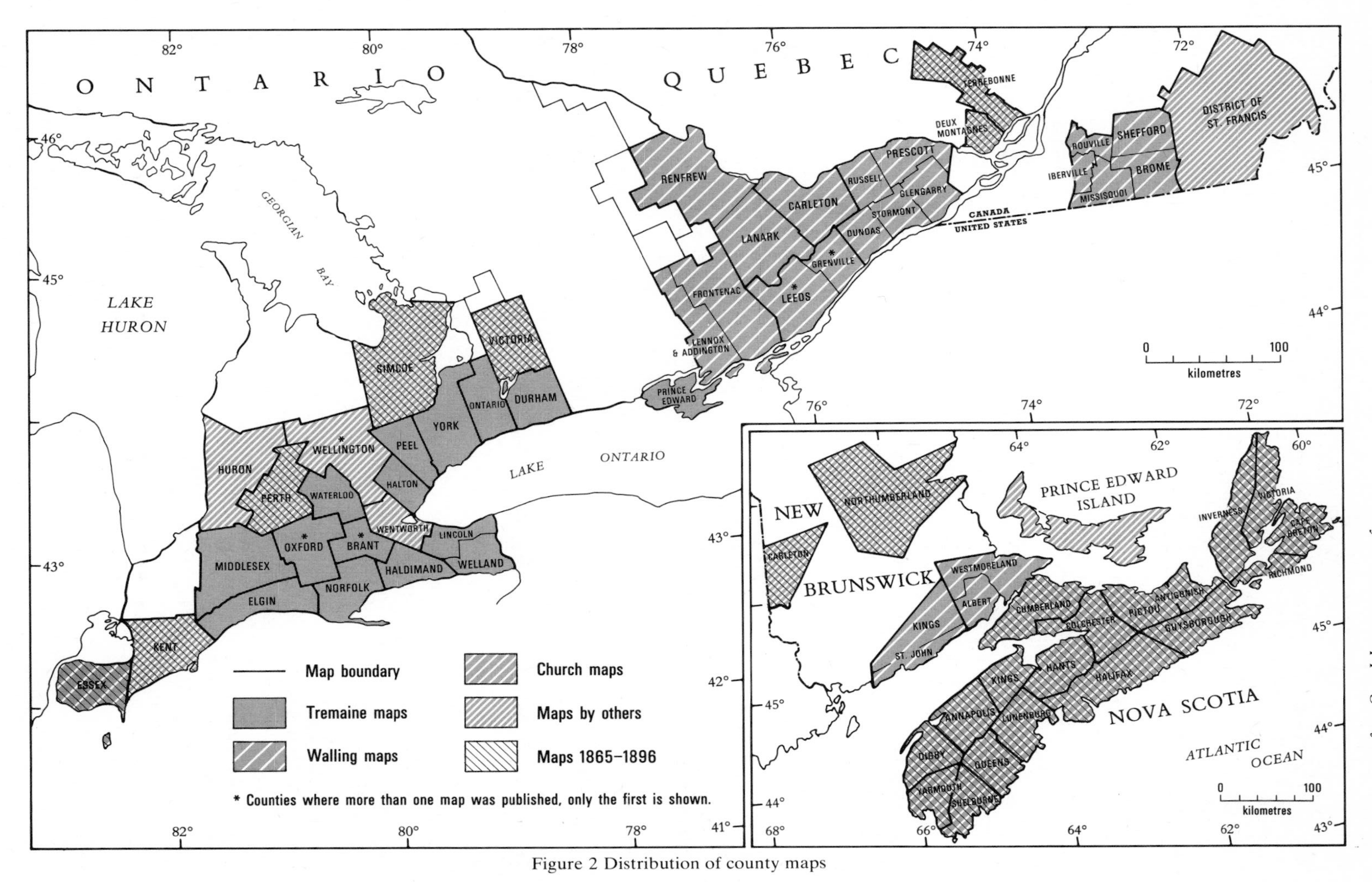

Figure 2 Distribution of county maps

desk and at the same time be a handy business reference was at once apparent, and the county atlas trade was on its way.

Although the basic framework of the county maps and atlases was provided by the township survey plans, the newer products were often more accurate than the original, and always had more up-to-date topographic and cultural information.[2] As has been mentioned, these products were sold by subscription and this necessitated a visit by the salesman to each of the landowners. It was the practice at the time of this visit to inquire about the dimensions of the holding, and (particularly if a subscription had been sold) to sketch in the farm buildings, ponds, copses, etc. This was done by eye within the dimensions of the survey pattern which was evident in the boundary fences. Most farmers knew that the original survey of the township had in some cases not been too accurately done, so many of them had privately remeasured their property to see if there would be a tax advantage in claiming less land than that shown on the official plan. Thus the county farmers knew the measurements of their land and were able to provide the county atlas representatives with data to make a more accurate map of the county than any previously available.

In many ways these maps showed more useful data than the topographic maps of today at equivalent scales. Not only were the land holders shown by name, but the town and rural industries were likewise identified. Churches were identified by religion, inns and taverns were shown with, in some cases, the parenthetical note 'temperance' for the information and convenience of lady travellers. All-in-all they were a most useful product in a young country where travel and transportation were difficult, and yet where a rising industrialization made local geographical knowledge important.

The distribution of county maps is shown in Figure 2,[3] and a listing giving the publisher, scale, and date of printing is provided in Table 1. In looking down the list of publishers one can see that some individuals, such as the members of the Tremaine family, H. F. Walling, and A. F. Church, made a career out of map publishing. Others, such as Lewes Bolton and J. H. Leclair, were essentially land surveyors and only published one or two maps of the immediate area where they practised their profession. This does not mean that the maps of the surveyors were better than those of the professional publishers, or vice versa. In point of fact, the publishers generally employed surveyors to compile their maps, while the surveyors, on the other hand, invariably had professional help in getting their maps lithographed or engraved.

County maps can from time to time be found in the stocks of map dealers. Most government and university map libraries have some of these maps, usually those of their own areas. The National Map Collection, Public Archives of Canada, has a complete set.

Table 2 County Atlases

Province and County	Date	Original publisher	Reprint publisher
Prince Edward Island King's, Queen's and Prince counties—	1880	J. H. Meacham & Co.	Mika
single volume	1925	Cummins Map Company	None
Nova Scotia			
Pictou	1879	J. H. Meacham & Co.	Mika

Table 2—*continued*

Province and County	Date	Original publisher	Reprint publisher
New Brunswick			
St John	1875	Roe and Colby	Mika—St John and York as single volume
York	1878	Halfpenny & Co.	
Quebec			
Eastern townships and southwestern Quebec	1881	H. Belden & Co.	Cumming
Ontario			
Brant	1875	Page & Smith	Cumming and Mika
Bruce	1880	H. Belden & Co.	Cumming
Carleton	1879	H. Belden & Co.	Cumming
Elgin	1877	H. R. Page & Co.	Cumming
Essex	1881	H. Belden & Co.	Cumming (with Kent)
Frotenac, Lennox and Addington	1878	J. H. Meacham & Co.	Mika
Grey	1880	H. Belden & Co.	Cumming
Haldimand	1879	H. R. Page & Co.	Cumming (with Norfolk)
Halton	1877	Walker & Miles	Cumming
Hastings and Prince Edward	1878	H. Belden & Co.	Mika
Huron	1879	H. Belden & Co.	Cumming and Mika
Kent	1881	H. Belden & Co.	Cumming (with Essex)
Lanark	1880	H. Belden & Co.	Cumming (with Renfrew)
Lincoln and Welland	1876	H. R. Page & Co.	Cumming
Middlesex	1878	H. R. Page & Co.	Cumming and Mika
Muskoka and Parry Sound	1879	H. R. Page & Co.	Cumming
Norfolk	1877	H. R. Page & Co.	Cumming (with Haldimand)
Northumberland and Durham	1878	H. Belden & Co.	Mika
Ontario	1877	J. H. Beers	Cumming and Mika
Oxford	1876	Walker & Miles	Cumming and Mika
Peel	1877	Walker & Miles	Cumming
Perth	1879	H. Belden & Co.	Cumming and Mika
Prescott and Russell	1881	H. Belden & Co.	Cumming (with Stormont, Dundas and Glengarry.)
Renfrew	1881	H. Belden & Co.	Cumming (with Lanark)
Simcoe	1881	H. Belden & Co.	Cumming
Stormont, Dundas and Glengarry	1879	H. Belden & Co.	Cumming (with Prescott and Russell)
Victoria	1881	H. Belden & Co.	None
Waterloo	1881	Parsell	Cumming (with Wellington)
Wellington	1877	Walker & Miles	Cumming (with Waterloo)
Wentworth	1875	Page & Smith	Cumming
	1903	Scarborough Company	None
York		Miles & Co.	Cumming and Mika

Note: addresses of reprint publishers: Cumming Atlas Reprints, Box 23, Stratford, Canada N5A 6S8; Mika Publishing Co., Box 536, Belleville, Canada K8N 5B2

County atlases are listed in Table 2.[4] Because books survive better than large maps, these atlases are often seen in rare book stores. In recent years a number of atlases have been reprinted, and the right-hand column of Table 2 shows the name of the reprint publisher.

Military Mapping 1863–71

Although the military mapping agency in Canada has through most of its existence contributed to the general topographic mapping of the country there have always been special maps produced for special military use. As early as 1758 we find reports of the military engineer, Samuel Holland, (who was later to become the Surveyor-General of Canada) drawing topographic maps during the siege of Louisbourg. After the fall of New France, General Murray, the first British governor of the colony, ordered a series of maps to be drawn of the settled areas. This series of 44 sheets at a scale of 1:24,000 shows the topography in detail but was not accurate in scale. It was not advanced beyond the manuscript stage.

The Fortification Surveys

The first series of accurate military maps to be drawn in Canada consisted of the sheets of the Fortification Surveys. These were large scale (1:2500) and covered the immediate vicinity of strategic points along the St Lawrence River at Quebec, Sorel, Montreal, Lachine, Vaudreuil and Kingston. The sheets of this series are the most detailed topographic maps that have ever been drawn in Canada. On them every tree, bush, shed, and hedge is shown. Even the backyard water pumps are depicted. Although each sheet only covers 1½ square miles (they measure 1 mile north–south and 1½ miles east–west) there were in all 197 sheets, so an appreciable amount of terrain is depicted around the six strategic centres. Most of the sheets held in the Public Archives of Canada are in black and white, but at the time they were produced (i.e. between 1863 and 1871) a certain number of copies of each sheet were coloured by hand. The following colours were used:

Buildings of stone or brick	Red
Wooden dwelling houses	Yellow
Sheds, barns and out buildings	Grey
Major roads	Buff
Open water (lakes, ponds, etc.)	Blue

Contours, shown by a black dot-and-dash line, were drawn at 10-foot intervals in the immediate vicinity of the town around which the sheets were drawn, and at 25-foot intervals further afield. In addition to the contouring, numerous spot heights are shown. All elevations are in feet above mean low water of the St Lawrence River at the town being mapped.

In each of the mapped areas the sheets of the 1:2500 series were reduced to six inches to one mile (1:10,560) and published as separate sheets covering in most cases four miles north–south and six miles east–west (i.e. the coverage of 16 1:2500 sheets). The six-inch maps were produced by simple photographic reduction, and as a consequence the lettering and fine detail is hard to read. Finally, index maps were drawn for each of the six areas. These indexes were at 1:25,000 for all centres except Kingston which has an index at 1:63,360. Table 3 gives the number of sheets at the three scales published for each area.

So far as can be determined, the sheets of the Fortification Surveys were never used for any purpose. Originally they were secret documents, and of course rightly so as they portrayed in great detail all possible approaches to the Canadian fortifications along the St Lawrence River. But even after the threat of an American invasion was long past these maps remained unused. Secret documents tend to remain locked in files much longer than necessary, and so it was with the Fortification Surveys.

Table 3 Fortification Surveys

Quebec and its environs (1865–67)	*Lachine and Caughnawaga (1867)*
Index map — 1:25,000	Index map — 1:25,000
Six-inch — 2 sheets	Six-inch — 3 sheets
1:2500 — 17 sheets	1:2500 — 25 sheets
Sorel (1867)	*Vaudreuil (1871)*
Index map — 1:25,000	Index map — 1:25,000
Six-inch — 2 sheets	Six-inch — 2 sheets
1:2500 — 27 sheets	1:2500 — 20 sheets
Environs of Montreal (1867–71)	*Kingston (1865–69)*
Index map — 1:25,000	Index map — 1:63,360
Six-inch — 3 sheets	Six-inch — 10 sheets
1:2500 — 34 sheets	1:2500 — 74 sheets

Military Sketch Maps

A kindred series were the military sketch maps. These maps were not produced by topographic surveyors but by regimental officers of the infantry and cavalry who had had a short course in the drawing of sketch maps. Most of this work took place in Canada during or immediately after the American Civil War at which time it was feared that the great armies of the Union forces, once they had subdued the South, might be turned against Canada. In the form of sketch mapping that was taught in those days, all available maps were examined and, if useful, were incorporated into the military map.[5] Thus the original township surveys formed the framework for these maps which were drawn of potential invasion areas such as the Niagara Peninsula and the eastern townships of Quebec (i.e. that part of the province lying between the St Lawrence River and the US border). Some of these sketch maps were not advanced beyond the manuscript stage, but most were reproduced by having the black detail printed by lithography and other colours (red for built-up areas, green for forests and blue for open water) added by hand. The scale was generally one inch to one mile though a few manuscript maps of the country lying along the north shore of Lake Ontario were drawn at two miles to one inch.

These maps, like the more detailed fortification surveys, had little if any use. One sheet, that of Fort Erie on the Niagara Peninsula, might have won fame as being the only military map produced in Canada in peacetime ever to have been used in battle. The battle (skirmish is probably a better term) was the repulse of the Fenian raiders of 1866. Unfortunately this map was not used because, evidently, the commander of the Canadian forces, Colonel Peacock, neglected to take the maps of the area with him when he marched his troops out of Toronto.[6] He was eventually required to deploy his force using a map torn from the back of Dewe's *Postal Guide.*

The Photo-Topographical Series at 1:40,000

This is a charming but relatively unimportant little series of 21 sheets surveyed and drawn between 1888[7] and 1892. Its description is included here first of all because it constitutes the first Canadian attempt at a truly topographic series, and secondly because the survey method used was a distinctly Canadian adaption of a system invented in Europe but little used there. In Canada ground photo-topography became the standard method of establishing mapping control in mountainous areas for almost 75 years.

Ground photo-topography is an extension of the plane table method with photographs replacing the topographer's view of the terrain through his alidade. As the camera is carefully levelled and carefully oriented along a specific bearing before the photo is taken, the resulting pictures can be used to obtain intersections of rays from two or more camera stations. The elevation of the intersected point can be calculated from measurements taken from the photos together with data on the position and elevation of the camera stations themselves as obtained by normal triangulation methods. With a good sprinkling of intersections (with their elevations) on the manuscript the topographer could then sketch in his contours. The use of photo-topography did not stop when this series was completed but continued into the twentieth century as one of the methods of obtaining control points for maps of the one-mile and four-mile series of the National Topographic System. After the Second World War, when vertical air photography and photogrammetric plotters became available, it was still used to provide height control for Multiplex plotter strips. The writer of these lines used the ground survey camera as late as 1951 in northern British Columbia.

Returning to the series under examination, each sheet covered an area 7½ minutes of latitude by 10 minutes of longitude or 8.7 miles north–south by 7.3 miles east–west. Five colours were used in the cartography. Black was used for cultural detail, the boundaries of the townships of the Dominion Land Survey system (see Chapter 3) and for border information. A green stipple was used to indicate forested areas. Blue was used for the surface hydrography with open water being shown by repeated shore lines, and glaciers by blue form lines. Contours and hill shading were in brown. Roads (in the few places they exist) and trails were shown in red.

The general appearance given by the sheets of this series is quite pleasing. Skilful hill shading compliments the contouring to give a good three-dimensional effect. The glaciers sketched with blue form lines, look cold and imposing. Rock drawing was not attempted so some of the ruggedness of the mountains is lost. Few place names are shown, possibly because few features were named at the time of survey. The township boundaries were not surveyed, but are shown in their theoretical position. To avoid the necessity of identifying the townships on the map a key to the township pattern is given in the bottom margin.

Before these sheets were published virtually all extensive survey operations in Canada were either cadastral surveys (where relief depiction was ignored completely or implied by rather primitive hachuring) or special engineering surveys for railways or canals. The notion of surveying contours to depict the terrain *per se* had either not been considered by the mapping authorities or had been dismissed out-of-hand as being much too expensive for a country the size of Canada. But when the land survey of the Prairies was extended into the mountains, the uselessness of a flat-land depiction was apparent to everyone. This series was an ingenious and economical solution to the problem of surveying mountain areas.

The production of the sheets listed in Table 4 and shown in Figure 3 was without a doubt a most useful exercise in both topographic surveying and mountain cartography. Due to the short period of time that this series was active, few sets of this series exist today. Complete sets are available for examination in the National Map Collection, and part sets are found in many of the older map libraries in western Canada.

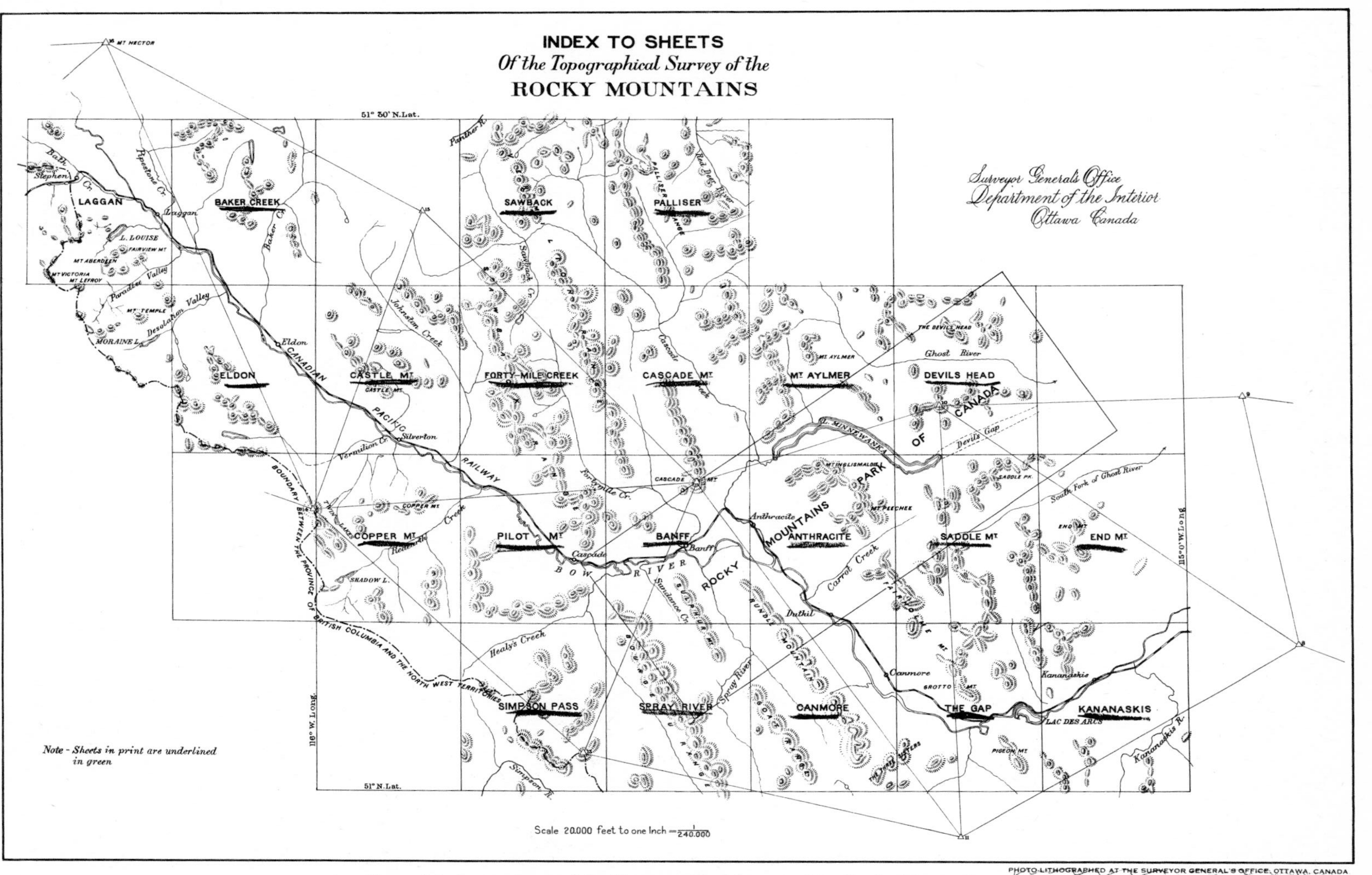

Figure 3 Index to sheets of the Topographical Survey of the Rocky Mountains

Table 4 Sheets of the 1:40,000 Series

Sheet name	Year of survey
Anthracite	1889
Baker Creek	1892
Banff	1888–89
Canmore	1890
Cascade Mountain	1890
Castle Mountain	1890
Copper Mountain	1891–92
Devil's Head	1889–91
Eldon	1892
End Mountain	1889–91
Forty-Mile Creek	1890
Kananaskis	1889
Laggan	1890
Mount Aylmer	1889–91
Palliser	1891–92
Pilot Mountain	1891–92
Saddle Mountain	1889
Sawback	1891–92
Simpson Pass	1891
Spray River	1888–91
The Gap	1890

3. The Three-mile Sectional Maps of the Canadian West

The three-mile maps (1:190,080) were the Dominion of Canada's first extensive map series. They were produced originally as a byproduct of the cadastral survey system that was set up in 1871 to provide farm lots for the settlers who went west onto the Prairies in the years following confederation. The original survey plans of the Dominion Lands Survey were at a scale of two inches to one mile (1:31,680), and although these were suitable for land registration it soon became apparent that a map at a smaller scale was needed for administration at the federal level. A six miles to one inch map (1:380,160) on which the six-mile square townships would appear as one-inch squares, was tried between 1884 and 1891 but this was found to be too small in scale. In the Annual Report of the Department of the Interior for 1892–93 the following statement is found:

> A record of all surveys made has hitherto been kept by compiling them on a scale of six miles to one inch on diagrams printed for the purpose. They exhibit at any time the state of the surveys of any part of the country. This scale has been found too small for the many miscellaneous surveys executed lately, and a change has been made to two miles to one inch. From these diagrams, maps at a scale of three miles to one inch are reproduced by photolithography; the progress of settlement is shown by indicating with three different tints the lands patented, those entered, and those reserved for various purposes.* Each sheet makes a map of convenient size, embracing a tract of land about 50 miles by 80. Five have been issued; they are Edmonton, Peace Hills, Calgary, Prince Albert North and Red Deer. Orders have been received to print an additional number of copies for the Dominion Lands agents and the public.

Thus was born the Three-Mile Sectional Map of the Canadian Prairies. Before this series was discontinued in 1967, 134 sheets were issued covering in all about 536,000 square miles (approx. 1,400,000 sq km).

* These tints were printed only on copies reserved for government use. Those sold to the public were black and white.

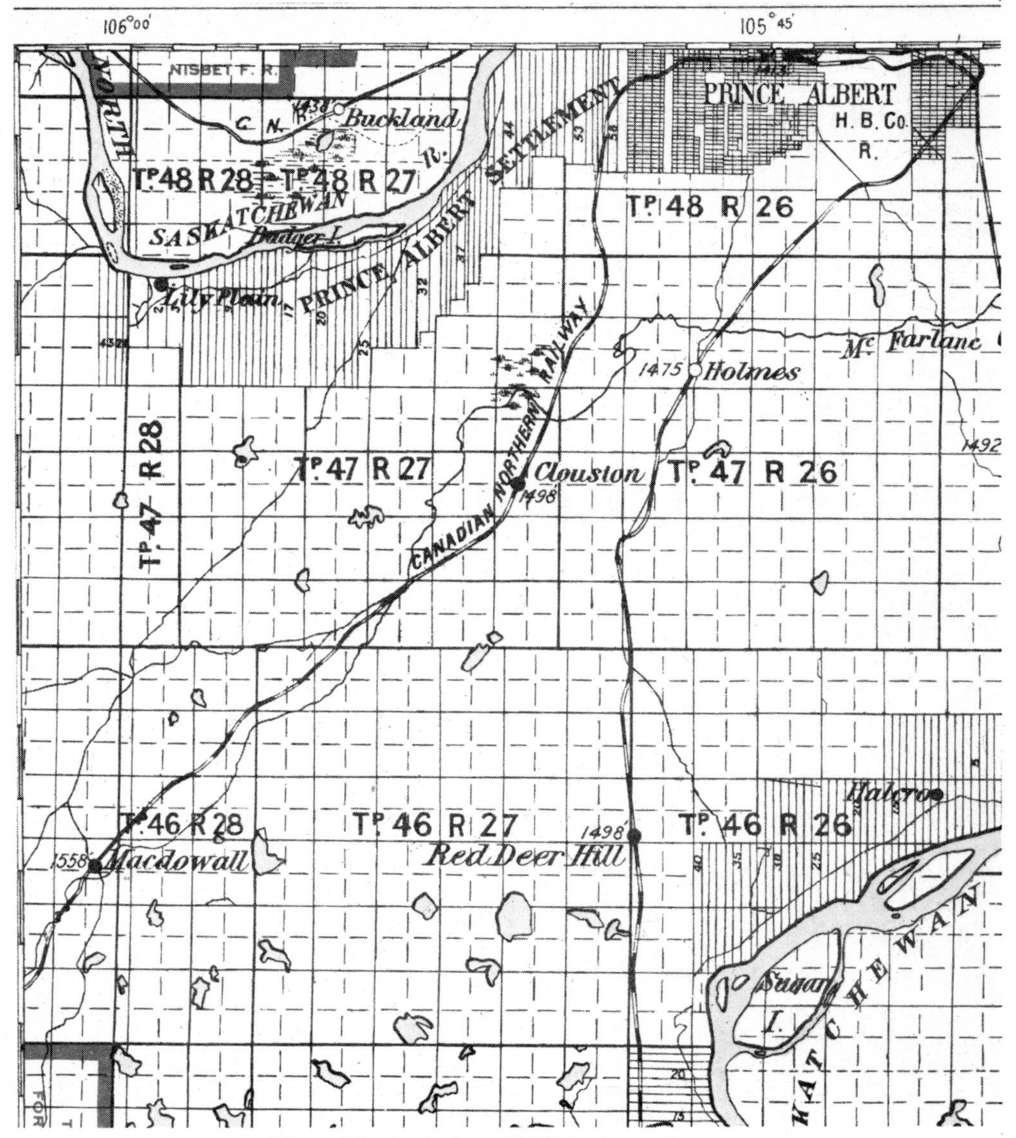

Figure 4 Section from an Old Style three-mile map

Styles within the Series

On examining any of the early sheets of this series one is struck by the spartan appearance of the cartography. The type used for the map title and number is a form black-letter type commonly used on posters. The remaining marginal information is in Roman but in point sizes that are far larger than necessary. The place names on the face of the map are somewhat less oppressive but still of a point size larger than that used on modern maps. The line work is coarse, with the lines of the land survey system dominating the whole presentation. In the original concept this series of maps was designed to give a simple presentation of the survey system and those topographic features that were necessary for

Table 5 Three-Mile Maps

Name*	Number**	Year of edition***	Final**** style
Port Moody	10	99, 02, 07, 13	O
Yale	11	99, 02, 05, 07, 10, 13, 21	O
Pincher Creek	14	96, 03, 05, 09, 12, 16	O
Lethbridge	15	94, 02, 07, 10, 13, 15, *24*	N
Milk River	16	02, 07, 10, 12, 14	O
Cypress	17	03, 07, 10, 12, 14	O
Wood Mountain	18	03, 06, 08, 11, 13, 16, *30*	I
Willowbunch	19	95, 02, 06, 08, 10, 12, 14, *30*	I
Souris (Weyburn)	20	94, 03, 06, 08, 10, 12, 14, 17, *24*	N
Turtle Mountain	21	94, 03, 08, 11, 13, 16, *22*	N
Dufferin	22	95, 02, 06, 09, 11, 13, 17, *22*	N
Emerson	23	95, 03, 06, 08, 11, 13, 17, *22*	N
Lake of the Woods	24	03, 06, 08, 12	O
Lytton	61	99, 03, 05, 07, 13, 17	O
Porcupine	64	96, 03, 08, 11, 12, 14	O
Macleod	65	94, 02, 06, 09, 11, 13, 16, *23*	N
Medicine Hat	66	95, 02, 07, 10, 12, 14, *24*, *47*	N
Maple Creek	67	96, 02, 07, 10, 12, 14, *25*	N
Swiftcurrent	68	96, 04, 07, 09, 11, 13, 16, *24*, *41*	N
Moosejaw	69	94, 03, 06, 08, 10, 12, 15, *21*	N
Moose Mountain	70	94, 06, 08, 10, 12, 15, *23*, *40*	N
Virden	71	94, 95, 02, 04, 06, 08, 11, 13, 16, *23*	N
Brandon	72	95, 03, 06, 08, 11, 13, 16, 16, *19*, *21*	N
Winnipeg	73	95, 03, 06, 09, 11, 13, 17, *21*	N
Cross Lake	74	03, 06, 12, 13, 26	O
Kamloops	111	99, 06, 07, 12, 12, 16	O
Sicamous	112	99, 03, 05, 12, 15	O
Spillimacheen	113	99, 03, 10, 14	O
Calgary	114	92, 03, 08, 12, 14, *20*, *26*, *55*	N
Blackfoot	115	94, 02, 06, 08, 09, 12, 15, *21*, *48*	N
Rainy Hills	116	96, 03, 06, 08, 10, 12, 14	O
Red Deer Forks	117	96, 03, 06, 08, 10, 12, 15	O
Rush Lake	118	96, 03, 06, 08, 10, 12, 14, *25*	N
Regina	119	94, 03, 06, 08, 11, 13, 15, *21*, *40*	N
Qu'Appelle	120	94, 02, 06, 09, 11, 13, 16, *25*, *50*	N
Riding Mountain	121	95, 97, 04, 06, 08, 11, 13, 15, 19	O
Manitoba House	122	94, 97, 06, 09, 11, 13, 17, 19	O
Fort Alexander	123	95, 06, 08, 11, 14, 21	O
Oiseau	124	15, 25	O
Seymour	162	99, 04, 10, 14	O
Donald	163	02, 04, 11, 13, *25*	N
Morley (Banff)	164	93, 97, 04, 07, 10, 12, 15, *25*	N
Rosebud	165	93, 97, 04, 06, 09, 11, 13, 16, *22*	N
Sounding Creek	166	96, 06, 08, 10, 12, 14, *32*	N
Bad Hills (Kindersley)	167	03, 06, 08, 10, 12, 15, *32*, *48*	N
The Elbow	168	96, 02, 05, 08, 10, 12, 15, *23*	N
Touchwood	169	96, 03, 05, 08, 10, 12, 14, *27*, *50*	N
Yorkton	170	94, 02, 06, 08, 11, 13, 16, *26*	N
Duck Mountain	171	94, 97, 02, 07, 10, 12, 15	O
Fairford	172	97, 04, 07, 10, 11, 14, 19	O
Washow	173	11, 14, 18	O
Athabaska	213	11, 14	O
Rocky Mountain House	214	93, 97, 04, 06, 08, 11, 13, 16	O
Red Deer	215	93, 94, 97, 04, 05, 08, 10 12, 15, *22*, *41*	N
Sullivan Lake	216	03, 05, 07, 09, 11, 13, 15, *26*	N
Tramping Lake	217	03, 05, 08, 10, 12, 14, *30*, *45*	N
Saskatoon	218	95, 02, 05, 08, 10, 12, 15, *20*, *27*, *40*	N

Table 5 Three-Mile Maps

Name*	Number**	Year of edition***	Final**** style
Humboldt	219	02, 05, 06, 08, 12, 14, *29*	N
Nut Mountain	220	02, 05, 07, 10, 13, 16	O
Swan River	221	00, 04, 08, 11, 14, 19, 46	O
Waterhen	222	12, 14, 18	O
Berens	223	17	O
Yellowhead	262	12, 16	O
Jasper	263	07, 10, 12, 14, 18	O
Brazeau	264	03, 07, 09, 12, 13, 16, 33, 48	O
Peace Hills	265	92, 94, 97, 03, 05, 08, 10, 12, 14, *21*, *48*	N
Ribstone Creek (Wainwright)	266	04, 05, 07, 09, 11, 13, 16, *24*	N
Battleford	267	94, 97, 04, 05, 08, 10, 12, 15, *28*	N
Carlton	268	94, 97, 04, 06, 08, 10, 12, 15, *35*, *48*	I
Prince Albert South	269	93, 97, 02, 05, 07, 08, 11, 13, 16	O
Pasquia	270	02, 06, 08, 13, 15, *23*, *46*	I
Mossy Portage	271	06, 08, 11, 14, *24*	I
Long Point	272	16	O
Brule	313	07, 10, 11, 13, 16	O
St. Ann	314	93, 97, 04, 07, 10, 12, 14, 17, 33	O
Edmonton	315	91, 94, 97, 03, 06, 09, 11, 13, 15, *20*, *28*, *40*, *47*	N
Vermilion	316	94, 97, 04, 06, 09, 12, 14, 18, *28*	N
Fort Pitt	317	97, 06, 10, 12, 14, 18	O
Shell River (Big River)	318	97, 04, 06, 08, 11, 14, *25*	I
Prince Albert North	319	92, 97, 02, 06, 08, 11, 14, 18	O
Carrot River	320	07, 11, 13, 16	O
Cedar Lake	321	14, 18	O
Grand Rapids	322	16	O
Simonette	362	15	O
Berland	363	13, 16	O
Fort Assiniboine	364	10, 12, 14, 17, *31*	N
Victoria	365	97, 04, 07, 08, 10, 12, 15, *29*	N
Saddle Lake	366	04, 08, 10, 12, 13, 16, *25*	I
Meadow Lake	367	12, 14, 18	O
Green Lake	368	12, 14, 18	O
Montreal Lake	369	14	O
Cumberland	370	15	O
Cowan River	371	14, 17	O
Minago	372	14	O
Wapiti	412	12, 14, 18	O
Losegun	413	13, 14, 17	O
Saulteux	414	13, 14, 22, *46*	I
Tawatinaw	415	10, 13, 14, 18, 48	O
La Biche	416	07, 10, 12, 14, 18	O
Primrose	417	15	O
La Plonge	418	14	O
Kississing	421	22	O
Wekukso	422	14, 18	O
Sipiwesk	423	14	O
Moberly	461	13, 17	O
Dunvegan	462	11, 14, 17, *28*	I
Smoky River	463	13, 15, *22*	I
Giroux	464	12, 14, 18	O
Pelican	465	13, 14, 18	O
Landels	466	13, 17	O
Dillon	467	19	O
Partridge Crop	473	14, *23*	I
St. John	511	13, 16, *22*	I

Table 5 Three-Mile Maps

Name*	Number**	Year of edition***	Final**** style
Montagneuse	512	13, 14, 18	O
Shaftesbury	513	12, 14, 15, 20, 48	O
Atikamik	514	15	O
Wabiskaw	515	14, 15	O
McMurray	516	13, 17	O
Methye	517	19	O
Limestone River	524	15	O
Notikewin	563	14, 18, 47	O
Penny River	564	16	O
Birth Hills	565	15	O
McKay	566	14, 18	O
Port Nelson	575	15	O
Wolverine	613	16, 19	O
Kokiu	614	16	O
Waskwei	615	16	O
Firebag	616	16	O
Mustus	663	14, 18	O
Mikkwa	664	14, 18	O
Lake Claire	665	21	O
Chipewyan	666	19	O
Dawson	1052	18	O

* Name in brackets is the final name given to the sheet.
** The missing numbers indicate that the sheet was never drawn.
*** All sheets were in Old Style for early editions; the italicized dates are those of the final style if the style was changed.
**** Style is indicated by O = Old, I = Intermediate, N = New.

the location of property boundaries, the whole to be shown in black and white at a convenient scale for use on the counters of the land registry offices. The poster-like appearance was probably intentional as no doubt the use of these sheets on the walls of railway stations, police posts and registry offices was commonplace.

The original sheets were revised frequently (as is shown in Table 5) particularly when cultural development, such as the building of railroads, made important changes in the landscape. As the population of the Prairies grew, these maps developed a wider readership. Press runs were increased, and eventually in 1912 a second colour (blue for open water fill) was used to give an improved appearance to the maps. In the following years an additional sophistication in cartographic presentation can be noticed developing. A third colour, green, was used to outline government land such as Indian reserves and forest reservations. In 1914 brown was used as a fourth colour to indicate relief, at first by hachures, but on certain sheets contour lines were attempted.

These developments were but a prelude to the complete change in cartographic style which occurred in 1920. During the previous year it had been decided by the authorities in the Department of the Interior that the sheets of the series should be changed into contoured topographic maps. This move would require new surveys to gather information on the additional detail to be shown, such as contours, dwellings, bridges and roads classified by colour.

Funds for an immediate and complete change-over were not available so selected sheets were earmarked for early conversion. Others were designated for a partial conversion into an 'Intermediate Style' which was produced by adding red and orange roads, brown contours, blue marsh and muskeg symbols to the Old Style base maps. None of the minor cultural detail such as houses, churches, bridges, etc. was to be shown on the Intermediate Style sheets. (These three styles were officially known as the Old Series, Intermediate Series and New Series, but here they will be called 'styles' to avoid the awkwardness of three 'series' within the overall three-mile series.) Table 5 shows the sheets that were converted to either the Intermediate or New Series and the year in which the conversion was made.

The Dominion Lands System of Survey

So many features on the three-mile maps depict, or are the result of, the Dominion Land Survey that it is almost impossible to discuss the series without first describing the essential features of the survey. The following is a short explanation of the principal features.[1]

TOWNSHIP. This is the basic unit of the survey of the Prairies being an area almost exactly six miles square situated with its sides lying on meridians and with its top and bottom being chords to parallels of latitude.

SECTION. Each township is divided into 36 sections, each section being one square mile. For purposes of land granting the section is divided into quarter sections. The quarter section ($\frac{1}{2}$ mile square or 160 acres) was the original homestead grant, though today farms and ranches of a whole section or more are quite common.

ROAD ALLOWANCE. An allowance for roads according to a set pattern was made during the survey of a township. Originally it provided for a strip 99 feet wide along all section lines but this was changed in 1881[2] to 66 feet wide on all section lines running north and south and on alternate section lines running east and west. In the settled areas almost all road allowances have been opened at least as dirt roads. Those that have been improved are shown by the road symbolization which will be described shortly.

CONTROL MERIDIANS. These are the north and south survey control lines that lie at intervals of approximately 4° of longitude across the Prairies. Thus the second meridian is at 102°, the third at 106° and so on to the sixth at 118°. The first meridian, called the *Principal Meridian*, was placed at a random point on the Prairies so as to avoid the riverine settlements on the Red River with no attempt, at the time, to set it on a given meridian. It was subsequently found to be at 97°27′28″ west of Greenwich. There is one control meridian to the east of the Principal Meridian. This is the 'Second Meridian East' at 94° W longitude.

BASE LINES. These are the east and west survey control lines run as six-mile chords to parallels of latitude. There is a base line every 24 miles (i.e. every four townships) north from the International Boundary.

CORRECTION LINES. The township sidelines are surveyed for 12 miles due north and 12 miles due south from each base line. It follows, therefore, that due to the convergence of meridians the townships will grow narrower to the north of the base line and correspondingly wider to the south of it. All townships touching the base line are exactly six miles wide at that point. This means that the township side lines do not meet those surveyed from the next base line to the north or to the south. As most Prairie roads follow the survey lines, this means that jogs in north–south roads occur along a line midway between base lines, such a line being called a correction line.

The Township Numbering System

The sections of each township are numbered from 1 to 36 in a zig-zag pattern starting at the south-east section. This pattern is illustrated by a diagram in the margin of all sheets, thus

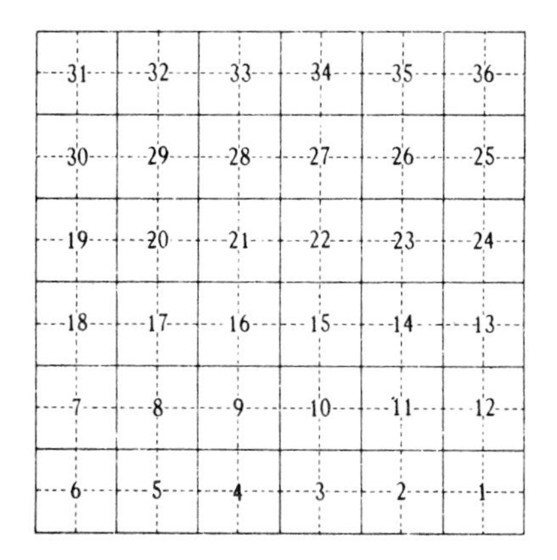

The columns of townships are called ranges and are numbered west from each control meridian. Within each range the townships are numbered north from the International Boundary. Each section is divided into four quarter-sections designated by the four quarters of the compass. This system is used to give the official description of land on the Prairies, and is written in short form, for example: NW$\frac{1}{4}$, 4, 21, 7, W3. This is translated to 'The north-west quarter of Section 4, Township 21, Range 7 west of the Third Meridian.'

The importance of this numbering system is shown by the prominence given to it on maps of the series. The township and range numbers were printed in bold type across the face of the map on all sheets until 1925. After this date the numbers were placed in the margins on sheets in the New and Intermediate Styles.

Sheet Coverage and Sheet Numbering System

The sheet layout is closely tied to the Dominion Land Survey system. As has been mentioned, control meridians were laid out on the ground every 4° of longitude. The sheets of the series were made to cover 2° in an east–west direction, and two of them, therefore, fit exactly between the control meridians. Thus a control meridian lies at either the east or west edge of each sheet. Each sheet covers from 13 to 15 townships in an east–west direction, the convergence of meridians causing the smaller number in the more northern areas. In the north–south direction, each sheet extends over eight townships, with a base line running along the bottom of the sheet, another across the centre and a third at the top of the sheet. The north–south dimension is therefore 48 miles or about 41′53″ of latitude.

Before 1914 the survey lines formed the neat line of each map, with the jogs at the two correction lines giving either the right or the left edge of each sheet a rather untidy appearance. After 1914 the coverage of each sheet was extended slightly in all directions so that the neat line could be made rectangular and a formal border could be put on each map. This produces an overlap with the four neighbouring sheets, but this overlap is small, never exceeding 2.5 cm east–west or 1 cm north–south.

Projection and Reference System

The projection used for sheets of this series is the simple conic with the central meridian at the centre of each sheet. In the early days no indication of latitude or longitude was shown but after 1914, when the sheet edges were squared off, divisions of one minute in both latitude and longitude were included in the formal border. (More details on the projection used for this series are given in Chapter 15.)

No rectangular grid system was ever used on sheets of this series, and, in fact, residents of the Prairie provinces have always been opposed to gridded maps at any scale. They have been quick to point out that they have their own point referencing system laid out right on the ground, and additional horizontal and vertical lines on the map only serve to complicate matters.

The Change in Style

The three-mile series started as a relatively simple map showing little more than the survey pattern, the drainage, the roads and railroads and the settlements. Colours were added to emphasize the presence of lakes and ponds or the boundaries of reserved lands. In fact, the only information found on sheets drawn in 1918 that was not present in the original design consists of the magnetic declination diagram and a slightly enlarged legend containing such innovations as a symbol for a telegraph line, spot-heights at railway stations and very occasionally contour lines which were printed in brown on a few of the Old Style sheets.

The change made in 1920 from the Old Style to the New Style and Intermediate Style was quite spectacular. During 1919, a complete study of the cartographic alternatives was made, a study which culminated in the publishing of a 'Change of Style' sheet which was drawn of a fictitious piece of terrain which included prairies, mountains, a portion of a sea coast and Edmonton representing a built-up area. The result of this study and experimentation was the cartographic presentation used on the New Style sheets. It is as light and pleasing as the Old Style was heavy and dull. The type on the map is small in point size but clean and easily read. Colours are used to advantage. The survey grid, while still omnipresent, is subdued, and made secondary to the road and drainage patterns.

The Intermediate Style was given the benefit of the above mentioned improvements in presentation but except for the inclusion of contours and a road classification, (differentiated by red and orange fill) very little information is shown that was not previously included on Old Style sheets. A comparison of the Old Style and Intermediate Style legends (See Figures 4 and 5) will bear this out. It is in the New Style, which will be described below, that the great change in topographic presentation took place.

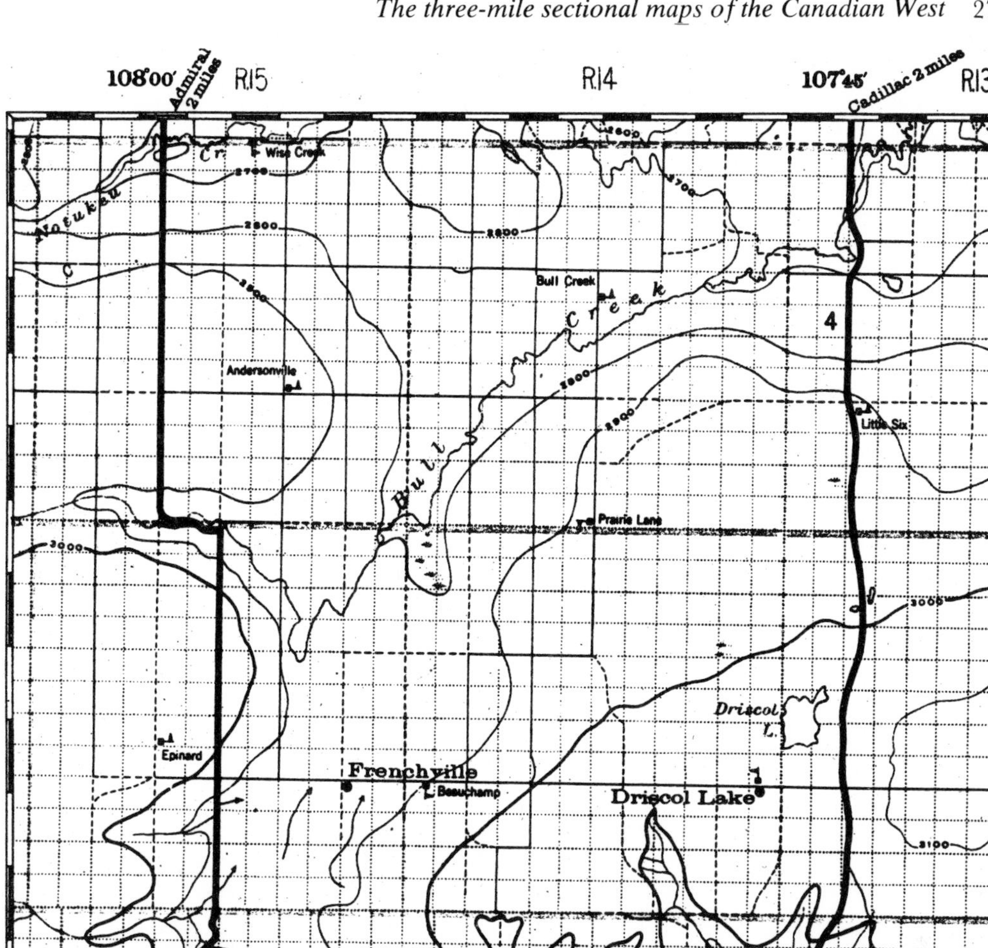

Figure 5 Section from an intermediate three-mile map

New Style sheets were drawn for the more highly developed districts on the Prairies. It must be remembered that all New Style sheets started out in the Old Style, and many went through several editions before being converted. In the whole of the three-mile series, 134 sheets were drawn, 39 of these were turned into New Style and 12 became Intermediate Style sheets. The remaining 83 stayed in the Old format throughout their existence. Figure 6 shows the distribution of the sheets in each style, and Table 5 lists the year of each edition and the style in which each edition was drawn.

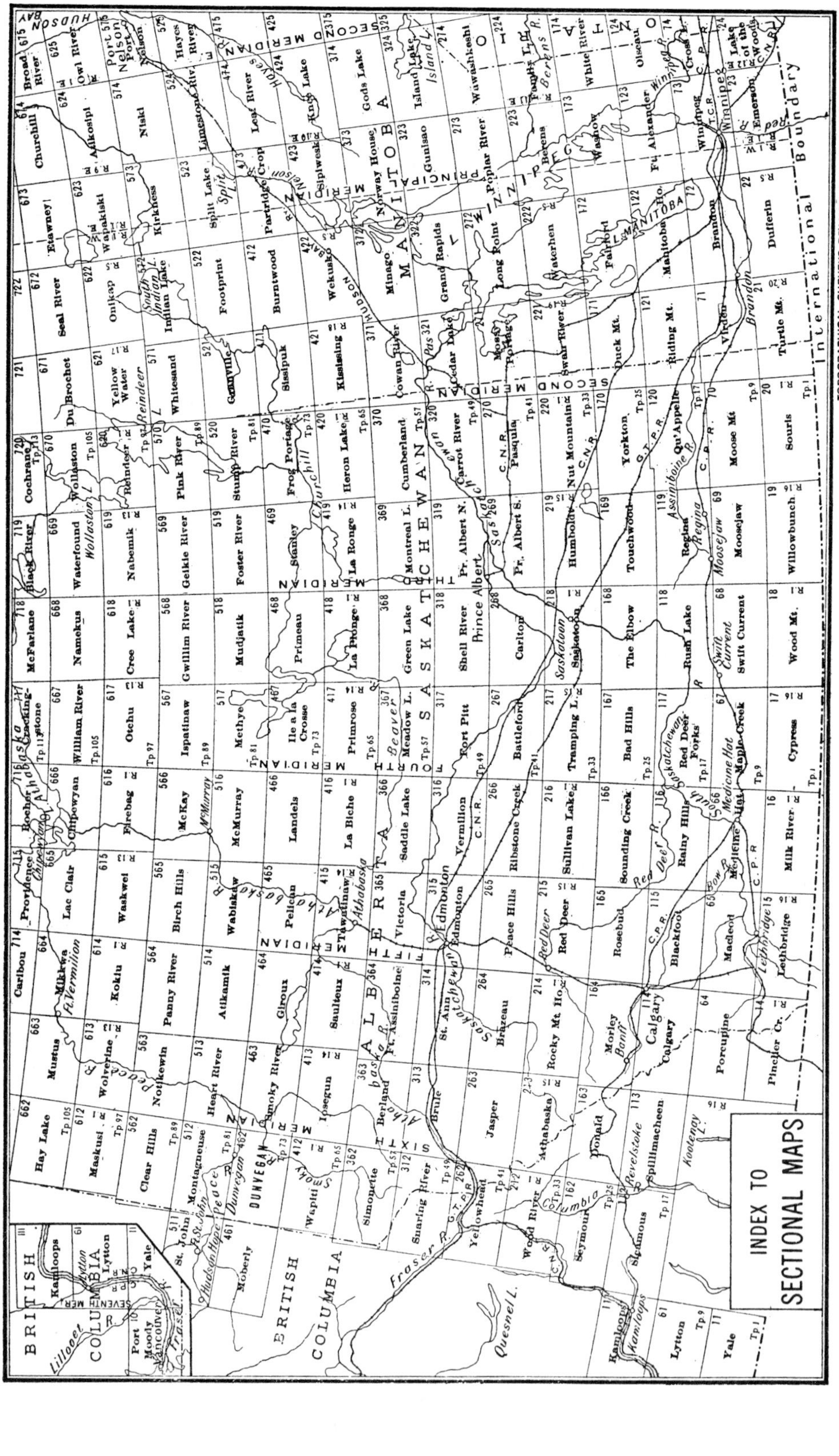
INDEX TO SECTIONAL MAPS
TOPOGRAPHICAL SURVEYS BRANCH, DEPARTMENT OF THE INTERIOR
BRITISH COLUMBIA
ALBERTA
SASKATCHEWAN
MANITOBA
International Boundary

Figure 6 Index to three-mile maps

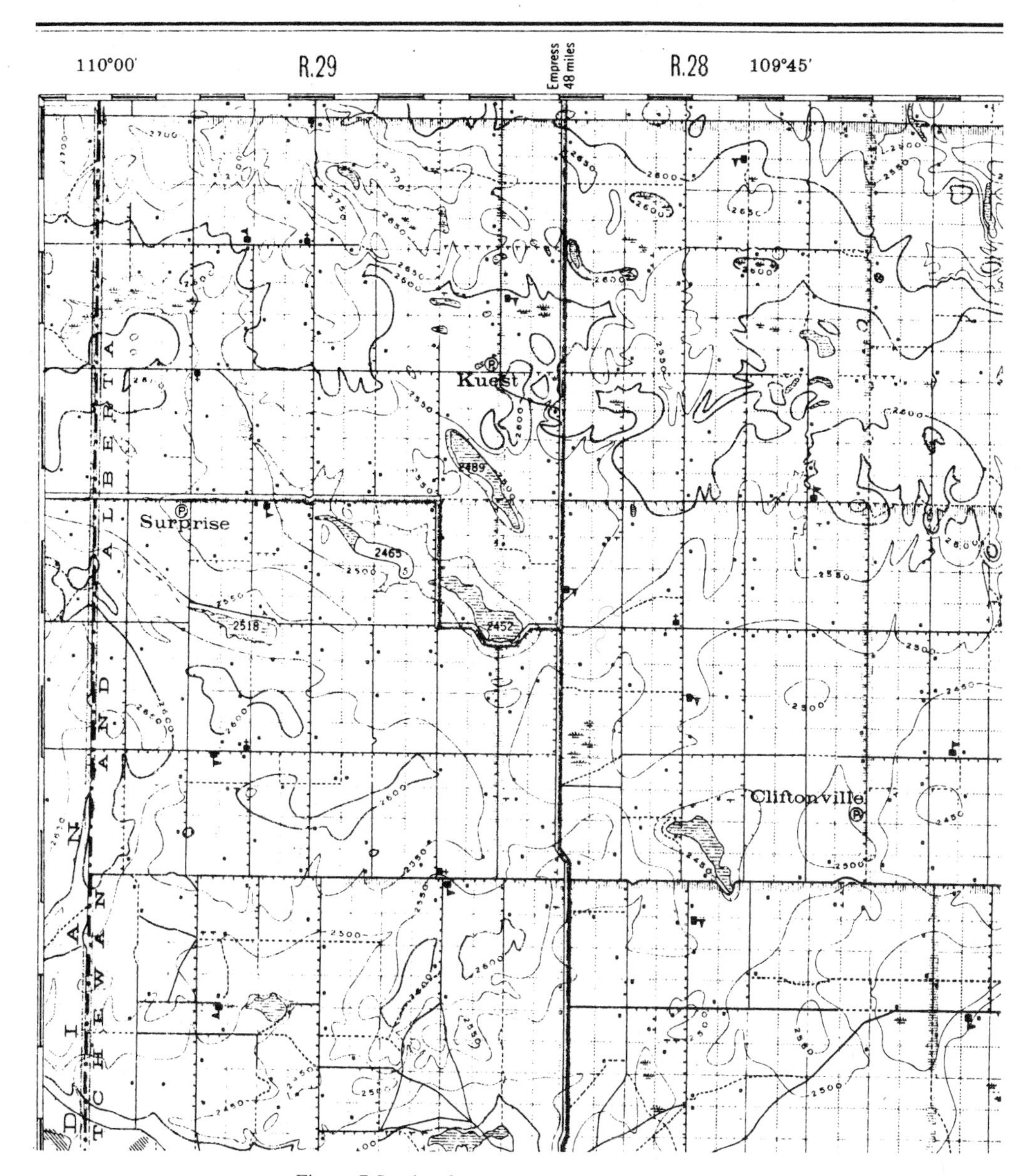

Figure 7 Section from a New Style three-mile map

New Style Sheets

A detailed description of the symbolization used on the New Style sheets will now be given. It must be remembered that between the change of style in 1920 and the last publication in 1955 there were minor modifications in colour and symbolization. The comments below refer to the presentation used on the majority of sheets.

Symbols and Colours

ROADS. Roads are classified in four different classes, depending on their condition and use:

Class 1	Main highways between cities are shown by a double black line filled in red. On some sheets the provincial highway number is shown in red.
Class 2	Secondary roads between towns are shown by a double black line left open on early editions but later filled with brown or orange.
Class 3	Local roads usually in good condition are shown by a solid black line.
Class 4	Unimproved roads or trails are shown by a pecked black line.

The New Style sheets and the automobile appeared on the Prairies at about the same time. For this reason these sheets were widely used as road maps. Up until the mid-1930s all gasoline stations were depicted either by a small red triangle or by the letter G. For a time this symbol competed with the symbol BS (Blacksmith's Shop), but eventually both symbols disappeared, the former because they became too numerous, the latter because they were too few.

RAILWAYS. It is no exaggeration to state that the railways were the moving force in the colonization of the Prairies. Without them there would have been no way of getting the produce of the grain farms and cattle ranches to the markets. Consequently, railways with their attendant features such as grain elevators, telegraph offices, sidings and stations have always been prominent features on the sheets of the three-mile series. The depiction on the New Style sheets is as follows:

Single-track line	a heavy solid black line.
Double-track line	a heavy solid black line with cross ties.
Electric railway	same as double-track railway but with finer line.
Stations	solid black oblong beside track.
Siding or stop	small black circle on line.
Telegraph office	the letter T in a small circle placed under the village name.
Elevator or elevators	the letter E in a small circle placed under the village name.

No attempt was made to place the T or E symbols at the correct position of the feature on the map. It was presumed that anyone on the Prairies would know that the telegraph office was at the station and the grain elevator would be beside the track nearby.

HYDROGRAPHY. All water features are shown in blue. Water features were of special interest to farmers and ranchers, especially in the south-west part of the Prairies where near-desert conditions exist. The differentiation between permanent, non-permanent and alkaline lakes was most important to settlers who would use such information as a basis for planning their agricultural operations. The water features are shown thus:

Permanent lake	light blue fill with dark blue outline.
Non-permanent lake	area filled with broken horizontal blue lines with outline in blue.
Alkaline flat	area filled with blue dots with outline shown by blue dots.
Marsh	area depicted by conventional blue bulrush symbol repeated over the area.
Muskeg	area filled with small coniferous tree symbol printed in blue.

Major river (over three chains wide)	double blue line with light blue fill between.
Minor river, stream, brook or creek	single blue line.
Non-perennial stream	broken blue line.
Irrigation canal	double straight blue line.
Drainage canal	single straight blue line.
Depth of lakes	blue dot followed by depth in feet in blue numbers. (The elevation of the water surface in feet is given in a black number.)
Glaciers	blue contour lines.

The names of all water features are in blue type.

FOREST. Wooded areas are shown by a green tint. Swamp areas (wooded marsh) are shown by combining the marsh symbol with the green tint. The classes of trees, whether coniferous or deciduous, are shown by black symbols scattered throughout the green-tinted area, a small pine-tree silhouette being used for coniferous areas and a bush-tree symbol (similar to the letter Q) being used for deciduous. On some of the southern Prairie sheets there is virtually no forest cover. Bluffs (i.e. rows of trees purposely grown as windbreaks) are shown as a strip of green in which a few tree symbols have been printed. Light green is used for areas of light woods, park lands and scrub. (Note: The term 'park land', like bluff, has a special meaning on the Prairies. Park land may be described as a naturally grassy area with a few widely separated trees.)

CITIES, TOWNS AND VILLAGES. The cities and towns are depicted by showing the street pattern with black rectangles filling the blocks in the built-up areas. When Class 1 or Class 2 roads pass through them, the routes are indicated. Smaller villages are shown simply by one or two building symbols together with the symbols for the facilities available in the village (railway station, garage, church, etc.).

BUILDINGS. Buildings are shown as black squares. In the case of farm buildings, one symbol only is shown for each group of buildings, except where there is more than one dwelling on one quarter-section. The public and commercial buildings (not already described) are shown thus:

School	small black square with flag.
Church	small black square with cross.
Post office	the letter P in small black circle. This is placed under the village name, not at the actual location.
Hotel	the letter H placed beside a building symbol.

TELEGRAPH, TELEPHONE AND POWER TRANSMISSION LINES. Telegraph lines preceded the railways into many parts of the Prairies and formed a valued connecting link with the outside world. Eventually they became a part of the rail network and were only shown on the map where they leave the railway and cut across country.

The telephone was in limited use in the area even before the first sheet of the series was drawn, but the position of rural telephone lines was not depicted until the New Style was adopted in 1920. For a time the location of telephone lines was an item of cartographic information of some value, but toward the end of the use of the series, telephone lines had become so common that an attempt to keep the symbol up-to-date was abandoned. Telegraph and telephone lines were both symbolized by a line of small T's where they ran across country. Along the railways they were not shown at all as it was common knowledge that telegraph lines were essential to the running of the railway. Along roads, telephone lines are indicated by small black ticks at right angles to the road. For the convenience of the cartographers these ticks were always shown on the south or west side of the road.

Power transmission lines appeared on maps drawn or revised after 1925. The symbol was originally a heavy dot followed by two dashes, but this was changed to a series of red inverted T's. These were also reduced to ticks where the line had been built along a road.

RESERVED LANDS. As these maps were drawn in the first instance as government administrative maps, the position of reserved land was probably given more prominence in the map design than would otherwise have been the case. The following outlines were used:

Forest reserves	purple band of colour.
Indian reserves	red hatching.
Dominion parks	orange band of colour.
Bird sanctuaries	grey hatching.
Public shooting grounds	blue hatching.

RELIEF OR 'LAY OF THE LAND'. The great advance over the Old Style sheets was the use of contours. It was estimated at the time that a survey party of from four to six men could carry out in one survey season (May to October) the additional surveys required to change an Old Style into a New Style sheet. Although all necessary additional information was gathered by the party (location of buildings, classification of roads, etc.) the main task was the carrying out of barometer surveys along the roads and the interpretation of the contours between the elevations so obtained. An automobile was provided for each party together with a list of railway elevations which would enable each track crossing to be used as a datum point. A special case that held three barometers was designed for this work, and at the time the barometers used were considered to be the most precise available. Fifty-foot contours were drawn of the Prairies. This was increased to a 100-foot interval in the foothills and to 250 feet in the mountains. By current standards this contouring would be rated as poor, but it was sufficient for the needs of the majority of map users of the series.

SURVEY INFORMATION. It was not the practice to indicate the horizontal control points of the Dominion Land Survey on maps of this series. During this survey an iron post was planted at each section corner and at the quarter-section points along section lines. These points formed the horizontal control for the series, and as they were so numerous (occurring every $\frac{1}{6}$ inch along section lines) it would have been pointless to show them on the map. However, a special symbol, a small black diamond, was used for special reference monuments such as those along the international or interprovincial boundaries. A small triangle indicates the position of the triangulation survey stations of the 1925 resurvey of the International Boundary.

Heights above sea level are shown by a black dot followed by black numbers indicating the height in feet. These were normally situated at railway stations, and the value given was for the top of the rail in front of the station. During the surveys of the bush country north of the Prairies, generally from about 1910 on, spirit level elevations were carried along the control meridians and the base lines. The elevation of the water level of streams cutting these lines and the elevations of township corners are shown on such sheets.

During the course of the survey operations, greatest care was taken in the survey of the control meridians, the base lines and the township outlines. On the map the meridians are named, the base lines are numbered north from the International Boundary, and the township outlines are emphasized by being shown in a grey hatched line on the map.

MARGINAL INFORMATION. Along the borders of the map the minutes of latitude and longitude are divided off, the value being shown every fifteen minutes. The township and range numbers of the survey system are either shown across the face of the map or, on New Style sheets, in the margin. Distances in miles to the nearest town or city are shown opposite first- and second-class roads at the point they leave the sheet.

All sheets have a diagram showing the numbering of sections within a township, and an index diagram showing the name and number of adjoining sheets. A third diagram showing magnetic declination was added to all sheets after about 1927. This last diagram had actually been introduced on some of the Old Style sheets as early as 1914 when the mapping reached the forested areas north of the plains. On the Prairies, with section lines at every hand, all laid out in the cardinal directions, there was no need for a compass. But in the northern forests it was common practice to survey only the township outlines and leave the section lines unsurveyed until settlers arrived who wanted to homestead in the township. Without the aid of section lines, the compass was essential for safe travel and consequently the declination information was included on the map.

The remaining marginal information consists of a good legend (called a reference in this series), bar scales in miles and kilometres, the map title, number, the reference meridian that lies to the east of the area mapped, and a credit note giving the date and previous mapping of the area shown. The practice of giving the edition number, which was faithfully followed on the Old Style sheets was not carried into the New Style.

The End of the Three-Mile Series

In 1925 an overall topographic mapping plan was devised for Canada which prescribed map scales in the geometric progression of 1, 2, 4, 8 and 16 miles to the inch. This National Topographic System, (NTS) as it is called, is described in detail in Chapter 15, so it is sufficient here to point out that the three-mile series was not part of it. Three-mile mapping was not immediately abandoned, but a stop was placed on plans to extend its coverage. Thus the last new sheet to be published was 421—Kississing, which appeared in 1922. As can be seen in Table 5, existing sheets were revised with commendable frequency through the years preceding and following the Second World War, and many Prairie map-users thought that this useful series would be allowed to continue as an alternative to NTS mapping. But this was not to be. In 1948 a definite 20-year programme for the completion of the four-mile series was formulated,[3] and the cartographic resources previously used to revise three-mile sheets were diverted into this new programme. Three sheets that were in work in 1948 were allowed to be printed. These were Qu'Appelle and Touchwood which appeared in 1950 and Calgary which lingered on until 1955.

In 1950 the conversion of the four-mile series to 1:250,000 began (see Chapter 7) and in 1953 the first three Prairies sheets of this new scale (Edmonton, Wainwright, and Winnipeg) were published. At this point the government sale of the three-mile maps was stopped when the area covered by a sheet was also portrayed on a 1:250,000 map. This policy continued until the replacement was completed in 1967 when all remaining three-mile maps were taken out of stock.

This series documented the settlement of the Prairies through the first 100 years of Canada's existence. In many cases the first indication a settler had of the location of his homestead was a pencilled mark on the three-mile sheet in the land agent's office. The series was used by farmers, ranchers, prospectors, Mounted Policemen, merchants, school teachers, in fact by anyone who travelled the west or had dealings with the people who lived there.

4. The One Inch to One Mile Series

The one-inch series and its successor the 1:50,000 are the most important series in Canadian mapping. The 1:50,000 scale is the largest scale at which large areas of Canada have been mapped, and it is the largest scale for which complete coverage of the country has been programmed. Over half the country is now mapped, and this includes all of the settled parts of the country and regions of the north that are important for resource development or national defence. The series when completed will include 13,150 sheets of which by August 1979, 7700 had been published. The Canadian one-inch series was converted to the 1:50,000 series by photographic enlargement in the years immediately following 1950. There was no change in style on the face of the maps, though, as will be described, the larger size of map requires some adjustment in format. The sheets of both series have from the first been actively used by many government departments, both federal and provincial, as base maps for thematic map series such as geological, forestry, land use, and so on. The principal use, however, remains that of a general purpose topographic map. From the point of view of the history of development the two series can be considered as one because the older series was merged completely into the newer. In fact, due to an uncertain policy on edition numbering at the time of conversion, many of the new 1:50,000 sheets appeared marked as second, third or fourth editions being considered simply as new editions of the old series.

The Beginnings of the One-Inch Series

In 1902 certain senior officers in the Department of Militia and Defence in Ottawa became concerned about the general lack of detailed maps of Canada that would be suitable for military operations. Experience in the Boer War had indicated the need for topographic maps, even in sparsely populated countries. Obviously some sort of mapping plan was needed, and to obtain advice on this a British military survey expert, Major E. H. Hills, was invited to Canada to review the situation. Major Hills arrived in Ottawa in May 1903 and immediately began examining the existing maps that were of scales suited to military planning. His study of Canadian topographic maps must have been short because in that year less than 1600 of the 3,500,000 square miles of the country had been so mapped. The work that had been done consisted of the photo-topographical work in the Rocky Mountains (see Chapter 2) and the maps of the Fortification Surveys at 1:2500 scale (also Chapter 2). Hills' report,[1] which was dated 31 December 1903, recommended that the Canadian Army create a topographical unit and immediately start mapping the strategic areas of Canada. It can only be presumed that this report was responsible for the formation, in 1904, of the Survey Branch in the Canadian Army's Intelligence Department. This unit started to work in the Niagara Peninsula, and in 1906 published two sheets, Niagara and Dunnville (now sheets 30 M/3 and 30 L/13 of the National Topographic System). The surveying and drawing of these maps were done in Canada but they were printed at the Ordnance Survey in England. The Ordnance Survey continued to print Canadian military maps until 1912 when the Survey Division (as the Survey Branch was renamed in 1906) obtained its own printing equipment. The difficulty in finding trained topographers in Canada led to an arrangement with the British Army for the loan of trained personnel during the summer months. This agreement started in April 1906 with the arrival of four sergeant-topographers who worked through the summer and returned to England in the autumn.[2] A similar group arrived each spring until 1914 when the threat of war in Europe brought an end to this help.

Long before Major Hills' arrival in Canada there was a topographical survey agency within the Department of the Interior. This unit worked under the Surveyor-General and although its main responsibility was the sub-division of the Prairies into farm lots, it had developed the photo-topographic method described in Chapter 2. After the completion of the 21 original sheets at 1:40,000 the unit continued to do some topographic work, mainly in federal parks and reserves. To give recognition of the ability of this unit to conduct topographic surveys, if asked to do so, it was given the title of Topographical Surveys Branch in 1890. In a memorandum written in 1920 Dr Deville, who held the post of Surveyor-General during this period, describes the style of mapping done by the Branch as follows: 'The nature of the information and the general style of the maps are much the same as for the British Ordnance Survey and the Survey of India.'[3] The third agency doing topographical work in Canada was the Geological Survey. The main mapping function of this agency was the provision of base maps on which geologists of the Survey could display the results of their field investigations. Due to the complete lack of large-scale mapping in most of Canada these maps were often pressed into service as general topographic maps. On occasion they were printed without the geological interpretation to facilitate their use as general maps. In 1908 it was decided to improve the quality of the topographic presentation by the surveyors of the Geological Survey. An American expert, Major R. H. Chapman, was brought in from the US Geological Survey to organize a topographic unit within the Geological Survey of Canada.[4] Chapman arrived in 1908 and immediately started to train the topographical surveyors in the latest American mapping methods. Arrangements were made to publish the topographic sheets separately as well as those with the geological information. To put the unit on a par with the other two topographic agencies it was given, in 1909, the title of Topographical Survey Division of the Geological Survey.

Thus from 1909 there were three agencies at work recording the topography of Canada:

1 The Topographical Survey Branch, Office of the Surveyor General, Department of the Interior.
2 The Survey Division, Department of Militia and Defence.
3 The Topographical Survey Division, Geological Survey, Department of Mines.

In 1922 the name of the Topographical Survey Branch was changed to the Topographical Survey of Canada, and in 1924 the Survey Division changed its name to the Geographical Section, General Staff (which was normally abbreviated to GSGS). For consistency these titles will be used throughout the description that follows even though the military unit changed its title twice again, once in 1946 to the Army Survey Establishment, and then in 1966 to the Mapping and Charting Establishment. Each of the three agencies had certain areas of primary responsibility and each had its own philosophy on map scales, sheet sizes and topographic specifications. It might be as well, here, to outline these areas of responsibility (as they were defined by the authorities of the day in the files of the departments concerned).[5]

Activities of the Geographical Section, General Staff

The military surveyors had as their area of primary interest the parts of southern Ontario and Quebec bordering on the United States. Today it is strange to think of the US as a possible invader of Canada, but in 1904 there were still employees in the Department of Militia who could remember the Fenian Raid of 1866. Even as late as 1903 a party of three Fenians attempted to blow up one of the locks of the Welland Canal, and although this last effort was handled as a police matter it did keep alive the possibility of military activity along the US border. In any event, this region and eventually parts of the Maritime Provinces became the areas first mapped by GSGS. The style of the mapping was strongly influenced by the British military topographers who in the early days worked alongside their Canadian counterparts; so the resemblance to Ordnance Survey maps was not a coincidence. The main scale used was one inch to one mile but some two-mile mapping was produced as will be described in the chapter on that scale. The one-inch sheets were 15 minutes of latitude and 30 minutes of longitude fitted into even degrees as illustrated in Chapter 15.

Activities of the Geological Survey

The Geological Survey sent its Topographical Division to regions where prospecting and mining activity was most pronounced. The topographers were trained in the American tradition, and like all topographers working closely with geologists they differed considerably from the military in what they considered should or should not be included on a topographic map. These differences will be examined more closely later in this chapter, but here two examples can be given. The first is the treatment of forested areas. Such features were not shown on the topographic maps produced by the topographers of the Geological Survey. To them the depiction of vegetation is both useless and a nuisance, as it obscures the geological symbolization without providing any data needed by their clients. Consequently none of the Geological Survey's topographic maps have any indication of vegetation.

The second example is the symbolization of the various classes of roads. Roads to a geologist are useful as landmarks and as possible routes of access to a given area, but they are not considered the vital arteries that the military consider them to be. Consequently the Geological Survey topographers would indicate the position of roads, but would not afford them an additional map colour to accentuate their position or to indicate the width and quality of the road-bed.

In the matter of the area covered by each map, the geologists preferred the 15 minute by 15 minute quadrangle used by the US Geological Survey, rather than the 15 minutes of latitude by 30 minutes of longitude employed by the GSGS. They did agree that the one-inch scale was the most useful, but at times they used two and four mile to the inch scales when the geological situation demanded such depiction.

Activities of the Topographical Survey

The overall plan of the Topographical Survey was to make a rapid survey of the Prairies at a scale of three miles to the inch (as was described in Chapter 3) and then, if these maps were well received, extend the coverage over the whole of Canada. At the same time a modest amount of mapping at the one mile and two miles to the inch scales would be carried out. The style of mapping, as has been mentioned, would be British and hence these maps would resemble those of the military.

Thus three separate topographic units emerged, two following the British tradition, the third the American. There was no coordination in the areas to be mapped and no attempt was made before 1922 to bring the various mapping styles together.

Efforts at Coordination

The complete lack of unity of purpose in the topographic mapping field became apparent far beyond the mapping agencies themselves. In 1892, 1908 and 1912, parliamentary commissions investigating government procedures were sharply critical of this lack of interagency cooperation.[6] Eventually the matter could not be ignored. In 1920 Dr Deville suggested, in a memorandum addressed to the departments concerned, that an interdepartmental committee or board should be formed to bring about some standardization and cooperation in Canadian mapping. This suggestion was accepted, and in 1922 the Board on Topographical Surveys and Maps was formed by Privy Council minute PC540 of 8 March 1922.

The Board only existed for three years, and in reviewing its activities it must be admitted that it had only modest success in achieving the coordinated mapping thrust that Deville so earnestly desired. Its major success was the development of the National Topographic Series (called, after 1950, the National Topographic System). This was, and is, the comprehensive mapping plan governing almost all Canadian topographic mapping in which the original set of scales were planned at one, two, four, eight and sixteen miles to the inch. The sheet lines of the smallest scale produced the basic NTS grid, (see Figures 43 and 44).

One of the fundamental decisions in the design of the NTS was the size of the basic one-inch sheet. The GSGS had been publishing 15 by 30 minute quandrangles for sixteen years and of course wanted to continue. The Topographical Survey agreed that this was the most convenient size, but the topographers of the Geological Survey wanted to retain the smaller 15 minute by 15 minute sheets. Consensus was finally achieved by agreeing that the Geological Survey could publish half sheets. Each half would bear the same sheet name, but would be distinguished by the addition of East Half or West Half to the name.[7] As will be seen in Chapter 5 the same device would be employed, 25 years later, in the early days of the 1:50,000 series. Full details on the NTS numbering system are covered in Chapter 15.

Agreement on the principal map scales must have caused some sadness in the Topographical Survey because their beloved three-mile series was not included. But the topographical and military surveyors entered into the spirit of cooperation and standardization. In fairness, it cannot be said that the Geological Survey did the same. In fact, in many instances they actively obstructed coordination. They steadfastly refused to adopt the 'British' style of mapping in their one-mile and two-mile series. They did not use the NTS sheet numbering system. They refused to cooperate in field survey projects. In this last matter some of the childish stratagems adopted by Dr Collins, the Director of the Geological Survey, are today almost beyond belief. He refused to tell Mr F. H. Peters (successor in 1924 to Dr Deville as Surveyor-General) where the geological field parties would be operating.[8] He refused to even consider the addition of a vegetation symbol[9] (of prime importance to the military) to the maps produced by his topographers, and on at least one occasion when the surveyors of the Topographical Survey found that they were working in an area that was adjacent to that surveyed previously by Collins' men, he refused to disclose the survey values of the monuments that had been established.[10] He was highly suspicious of the move in 1922 to change the name of the Topographical Surveys Branch to the Topographical Survey of Canada.[11]

The Board of Topographical Surveys and Maps held its last meeting on 14 April 1925. In September of that year its name was changed to the Board of Topographical and Aerial Surveys and Maps by Privy Council minute 1394½ of 1 September 1925. This action took into account the growing importance of air photography in topographic mapping, and sought interagency cooperation and standardization by organizing a central committee to consider both the standardization of maps and the acquisition of aerial photography for the mapping. Not too much time elapsed before the authorities realized that the development of techniques for plotting topographic maps from aerial photographs was changing the whole aspect of topographic mapping, and that the Committee's work in coordinating the provision of aerial photography was occupying virtually all of its time. To bring the terms of reference of the Committee into line with its major preoccupation the Committee was reorganized once again (by PC 1061 of 2 June 1933), and renamed the Interdepartmental Committee on Air Surveys and Base Maps. The stated duties of this Committee were 'to receive from all government services requests for air surveys and base maps and to lay out a programme giving priority to the most important work'. The PC minute goes on to say that the Topographical Surveys Branch of the Department of the Interior would henceforth be called the Topographical and Air Survey Bureau, and would be the 'central air surveying unit for the Dominion of Canada'. Unfortunately for those who looked for a strong unifying force in this Privy Council minute there were two exceptions to this centralized work. The exceptions were mapping for special military and geological work.

These enormous loopholes left matters pretty much as they had been, because the GSGS and the Geological Survey immediately pointed out that all the work they were doing was special to their needs.

In examining the maps produced after the PC minute was published it can be seen that the military authorities made an honest effort to come as close as possible to the Topographical Survey style. The topographers of the Geological Survey held rigidly to their concepts of topographic mapping.

The resistance of the Geological Survey to cooperate contained the seeds of its own downfall as an independent topographical survey unit. By 1935 it was apparent to many, both inside and outside the survey services, that amalgamation of the civilian topographic units must take place. This came about in 1936 during a widespread reorganization of government services. A new Department of Mines and Resources was formed which had within its structure the Bureau of Geology and Topography. The Bureau was composed of the Geological Survey of Canada (without its topographic division) and the Topographical Survey of Canada (which included both civilian topographic surveys). As with any amalgamation, some careers were advanced while others were retarded. The bitterness of the less fortunate was felt within the Topographical Survey well into the 1950s. The style of the one-inch maps that were produced from 1936 on was the British style, though of course this description is superficial as much of the map content is Canadian.

The American-style maps of the Geological Survey that had been produced before the amalgamation were not discarded, but were up-dated periodically by the correction of map detail caused by changes in the topography. Most have now been replaced by standard 1:50,000 sheets, but as late as August 1978 fourteen still remained in the system. All have been enlarged to 1:50,000 but they still are without forest symbolization and have uncoloured roads. No doubt they will be replaced within the next two or three years, but at the time of writing they constitute the only large-scale coverage of the quadrangles they occupy.

After 1936 the amalgamated civilian agency, which took the name the Topographical Survey of Canada, cooperated fully with the GSGS to design a standard format and develop a common set of specifications. From then to the end of the series, fourteen years later, the sheets of the two agencies were almost indistinguishable. Today if one examines the succession of editions of the maps of the three agencies one is struck by the way that the military design persisted and eventually emerged as the standard Canadian one-inch map.

The Introduction of Photogrammetry

In carrying out a programme of large-scale mapping (and in a country the size of Canada the one-mile scale must be considered large scale), the mapping authorities are always faced with the problem of producing, with available resources, the maps that are most urgently needed. Those planning the one-mile mapping were quick to recognize that aerial photography was a valuable cartographic tool, particularly in forested areas where traditional plane-table methods could not be used. The photogrammetric plotters developed in Europe were not used to any great extent in the one-mile mapping. There was only one plotter in Canada before World War II (it was owned by the GSGS) and the Topographical Survey did not purchase any until 1948. But once introduced the photogrammetric plotting instruments, almost within a year, put an end to the plane table.

The introduction of photogrammetry to the Topographical Survey came at the end of the life of the one-mile series. Even as the Multiplex plotters were being set up in the Topographical Survey offices in Ottawa, a meeting was being held in England at which representatives of the military mapping agencies of the United Kingdom, the United States and Canada were deciding the future mapping needs of the armies they represented. The desirability of switching from the imperial scales to the even representative fractions of 1:25,000, 1:50,000 and 1:250,000 was pointed out. The existence of almost complete coverage of Europe (always a possible battlefield) at these scales was a strong factor, and so also was the desirability of standardizing weapons, range finders and other equipment that is sensitive to distance measurement. Although not obliged, Canada made the change from the one-mile scale to 1:50,000 in 1950.[12] The reason was not simply to please the military. The realization even in 1950, that Canada would one day convert to the metric system swung the balance. All sheets going into work in 1950, that would have been drawn at the one-inch scale, were switched to 1:50,000, but those sheets already in the system were allowed to continue through at the inch to the mile scale. Thus the last map drawn at the old scale was 93 A/5 Beaver Creek which was published in 1952. Of course the one-mile sheets did not instantly disappear. Despite an heroic effort in 1951 and 1952 to convert as many as possible to 1:50,000 by photographic enlargement, some sheets of isolated areas remained in the Canada Map Office for many years. In August 1978 seven were still there.

Symbols and Colours

Attempts by the Board to standardize symbols and colours on Topographical Surveys and Maps was only partly successful. As has been mentioned, the GSGS and the Topographical Survey had much the same tradition in that both took their styles of mapping from the Ordnance Survey in Great Britain, and both had the same aims and objectives, namely to produce large-and medium-scale topographic maps. What is more important, neither organization considered the other a threat to its existence. Consequently there was a commendable standardization of product between these agencies, both by formal decisions and informal agreements. The Geological Survey found it more difficult to cooperate, but before too much scorn is directed toward that organization, it must be remembered that the topographic needs of the geologist are quite different from those of the general public, and in many of the isolated regions of Canada the geologist and prospector constitute the 'general public'. So in tailoring the topographic presentation to the 'regional' primary use, the Geological Survey felt it was best serving the public. The result was that in reality two separate Canadian one-inch sub-series emerged. Faced with this it is more convenient here to examine the symbolization and style of each sub-series separately. The one resulting from the combined work of the Military and Topographical Survey will be examined first.

Sheets of the Military and Topographical Survey[13]

ROADS. Road depiction in this series commenced before the widespread use of motor vehicles and hence the original road hierarchy was quite simple:

Metalled road	Two fine parallel black lines filled with contour brown on the military maps, red on the Topographical Survey maps.
Unmetalled or wagon road	Two parallel black lines without fill on the military maps, parallel black pecked lines on the Topographical Survey maps.

In remote areas both agencies used a single black pecked line to indicate a trail or portage. By 1927 paved rural roads were common and a new category was added. The Military Survey joined Topographical Survey in using red for the fill of the top category, the paved road. A gravel road was shown by the Military Survey by an intermittent red fill, and by the Topographical Survey as a brown fill. The lesser categories remained the same.

RAILWAYS. Both the Military and Topographical Survey used the same railway symbolization. Single-track lines were shown as a single black line with cross ties (sleepers). Double-track lines were depicted as parallel lines with cross ties going across both lines. Stations were shown as small rectangles—black in both agencies if the building was of wood, red on the military maps if the building was of brick or stone. Grain elevators, round houses and other railway accessories were shown as large buildings and were labelled.

HYDROGRAPHY. Both agencies used the same symbols for water features but minor variations occurred from time to time:

Permanent lake	Dark blue outline with either light blue fill or dark blue repeated shore lines (ripple lines). After 1927 the light blue fill was almost always used. The depths of lakes are not shown after 1936 but prior to that the Topographical Survey showed lake depths by a blue figure on the lake surface. The surface elevation of lakes was shown in feet above sea-level by black figures on Topographical Survey maps and brown figures on those of the GSGS.
Non-permanent lake	Area filled with pecked horizontal blue lines on sheets of both agencies.
Marsh	Area depicted by conventional blue bulrush symbol repeated over the area.
Swamp	Same as marsh but with forest symbol over the area in addition to marsh symbol.
Major river	Shorelines in dark blue with light blue fill or ripple lines. Rapids are shown by an interruption of the blue fill and short blue lines parallel to the shore at the site of the rapids. Falls are depicted by a blue line across the river at the site of the falls with short lines parallel to the shore pointing down stream from the fall-line.
Minor river, creek, stream or brook	Single blue line.
Non-perennial stream	Pecked blue line.
Glaciers	Blue form lines over ice area.

The names of water features are shown in black type on the Military Survey maps prior to 1927 but were changed to blue as a standardization measure on new sheets drawn after that year. The work required to separate the blue type from the black type was considerable and as a consequence many Military Survey sheets retained black water type throughout the life of the series.

FOREST. Originally there were two treatments of the symbolization of forest cover. The GSGS was most complete in that it had different symbols for coniferous and deciduous growth. As the symbols were drawn individually by hand they were spaced out for light growth and close together for heavy. Mixed forest was indicated, logically, by mixing the symbols. Originally the Topographical Survey used a light and a dark green tint to indicate forest cover of light or heavy growth, but after 1936 adopted the Military Survey symbols. Muskeg—that ultra-Canadian vegetal phenomenon—is depicted in Topographical Survey sheets as an array of tiny blue coniferous trees interspersed with blue horizontal lines indicating the wetness of the ground. The symbol does not appear to have been used by the army, possibly because their mapping activities did not take them into muskeg country.

BUILDINGS. Probably the most striking feature of the early Canadian military maps was the separation of masonry buildings, which were symbolized in red, from those of wood which were shown in black. Such an elaborate classification of structures was quite beyond the needs of the Topographical Survey, and one suspects that even within the army unit it was by 1927 becoming a nuisance to maintain. In any event it was given up in 1927 in the name of standardization. Both the Military and Topographical Survey were careful to show all dwellings, sheds, rural industries such as saw mills, cheese factories, etc. On some park and Indian Reserve maps Topographical Survey indicated the position of isolated graves with a small black cross. This was dropped during the 1930s as being an unnecessarily sombre aspect of cartography. It is interesting to note that from 1936 on, Topographical Survey used the army classification of bridges into iron, wood, concrete or stone construction.[13a] Although it is not mentioned in the records of the time this must have been a move towards standardization because the army placed considerable importance on such information.

CITIES AND TOWNS. At the one-mile scale the block pattern of cities and towns can easily be shown, and both agencies did so. The early army cartographers used the red and black building classification for the centre of cities. Where buildings were too close to be shown individually, a pink tint was used to indicate that most structures were of masonry construction. Thus the pre-1927 built-up area treatment is much the same as that used today on the 1:50,000 series. After 1927 both the GSGS and Topographic Survey used a stylized block outline (in black) for city and town centres.

TELEPHONE, TELEGRAPH AND POWER TRANSMISSION LINES. These features were important landmarks and also are indicators of industrialization and resource development. Both agencies used red for power lines and both differentiated between *main lines* which were shown as miniature lengths of line hanging in catenary between tiny uprights, and *other lines* which were a row of small red T's. Telephone lines were shown as small black T's. Of course during the history of this series telephone lines became so common that they ran along almost all country roads. Later military maps specified that the symbol was used only for 'trunk routes'. Telephone and telegraph offices were both identified by the letter T, the telegraph office being of slightly heavier type.

RELIEF. Relief was shown by both agencies by brown contour lines usually at 25, 50 and 100-foot vertical intervals depending on the nature of the country. Index contours, drawn in a heavier line weight are at every fourth contour for the 25-foot and 50-foot interval maps, but every fifth contour for those with a 100-foot interval. Bench marks are shown by the broad arrow with the elevation to the nearest tenth of a foot written nearby. Spot elevations were shown by both agencies, before 1936, by a brown saw buck (small x) with the elevation to the nearest foot nearby. Both agencies used a black dot plus the elevation after that year. Contour numbers are placed to read upright from the bottom of the map and do not always read up-hill. The European custom of placing the numbers so that the top of the number is on the higher ground would later be adopted in the 1:50,000 series.

SURVEY AND CADASTRAL INFORMATION. Triangulation stations are indicated by small triangles with a centered dot. Survey lines are given various treatment. In the Dominion Land Survey part of the country (i.e. the Prairies and southern British Columbia) most survey lines are marked on the ground by a road or trail and no further symbolization is necessary. In wilderness areas survey lines are shown by the trail symbol (pecked black line) but can be distinguished from trails by their straightness and a north–south or east–west orientation. There does not seem to have been a consistent policy for the indication of the cadastral framework in the non-DLS parts of the country. On some of the early army maps the concessions and lot numbers (but not the lot lines) are carefully shown. On later editions of the same map this useful information was taken off. There was certainly no cartographic reason for not showing the pattern of lot numbering as this was quite effectively accomplished on the Chief Geographer's maps at 1:250,000 scale (see Chapter 10), but it does not seem to have been required during the last years of the one-inch series.

MARGINAL INFORMATION. Almost a complete legend was used by both agencies. It was divided into two parts and placed in the bottom margin with one part on each side of the title block. Under the sheet title one finds the province or provinces covered by the map, the scale, and the contour interval. The scale is given in three ways, a verbal statement (sometimes as 'one mile to one inch' but just as often as 'one inch to one mile'), the representative fraction 1/63,360 and as bar scales showing miles and thousands of yards. An index map showing the sheet location and the location of any other sheets in the vicinity is placed at the right side of the bottom margin. At the extreme left is the imprint note giving the name of the producing agency, the dates of the survey and of any revisions, and the magnetic declination. Nothing is placed in the side margins, and the top margin contains only the name of the producing agency, the words National Topographic Series, the edition number and the sheet number. Although this may sound as though the marginal information was rigorously standardized, this was not so in practice. The minor variations were so numerous it is difficult to find two sheets with identical marginal information. One completely innovative issue was the 1944 edition of the Lachine Quebec Sheet 31 H/5 which has all the marginal information in both English and French. This style was not extended to other provinces during the life of the one-mile series but was adopted for the 1:50,000 series in 1960.

The map border used by the Military and Topographical Survey was quite elaborate. Immediately next to the neat-line is the latitude and longitude bar consisting of two parallel black lines one millimetre apart which contain alternating black and white minute bars around the map. About one centimetre outside this, are two more parallel black lines, these being blocked off in units of the grid advocated by the publishing agency (i.e. a 1000-metre Modified British Grid on the Military Survey maps and a four-mile Transverse Mercator grid on those of the Topographical Survey). The border is completed by a heavy black line one millimetre in thickness.

Sheets of the Topographical Division of the Geological Survey of Canada

As has been mentioned, although the topographical sheets produced by the Geological Survey are considered part of the National Topographic Series, they differ so much from those of the other producing agencies that they are essentially a separate series. The sheets so produced are part of the Geological Survey A Series and they all date before the amalgamation of the civilian topographical surveys in 1936. The A Series was initiated in 1910 to indicate a map published separately (i.e. not illustrating a particular geological memoir). Hence all the Geological Survey topographic maps were part of the A Series.

ROADS. Three classes of roads are used, namely, road, road not well travelled, and trail. The first uses two parallel black lines, the second parallel pecked black lines and the third a single pecked black line. No colour fill is used.

DRAINAGE. Rivers and lakes are depicted in much the same way as on the sheets of the other two agencies. On some sheets, to save on printing costs, rivers, lake edges, swamps, etc. are printed in black, but a blue fill is used to indicate open water.

RELIEF is shown by brown contours in the same fashion as on sheets of the other agencies. Contour numbers do not always read up-hill.

VEGETATION is not shown.

BUILDINGS. Less emphasis is given to public buildings and business enterprises but great attention is paid to mining and quarrying activity. Separate symbols are used for bore-holes, mine tunnels; mine tramways are distinguished from standard gauge railways and prospect workings are identified.

SURVEY AND CADASTRAL INFORMATION. This information (i.e. concession and lot numbers, survey lines, etc.) is of interest to prospectors and developers and is clearly shown.

SHEET SIZE. The normal size of sheet was 15 minutes by 15 minutes (i.e. half the size of the sheets of the other agencies). At times full sheets and sheets at odd sizes were drawn if the size and shape of the geological formations being mapped demanded such treatment.

MARGINAL INFORMATION. On the older sheets the legend is shown in the right margin in a column of small boxes in the American style. After 1932 the legend was moved to the bottom margin and resembled more the legend presentation of the other agencies. The NTS numbering system is not used but each sheet is given a serial number followed by the letter A.

Rate of Production

At the time that Dr Deville proposed the formation of the Board on Topographical Surveys and Maps (1920) each of the three agencies had had considerable experience in mapping at the one-inch scale. Estimates made at the time[14] credit each as follows:

Agency	Area mapped		Principal area of operation
GSGS	23,400 square miles	(55 sheets*)	Ontario and Quebec
Geological	12,840 square miles	(30 sheets)	Mostly British Columbia
Topographical	16,000 square miles	(37 sheets)	Mostly Alberta and British Columbia

* As many Geological and Interior sheets were irregular in size the figure given is for the equivalent number of 15 minute by 30 minute quadrangles of 425 square miles.

By 1939 the GSGS and Topographical Survey were both working on standardized sheets of the National Topographic Series while the Geological Survey had dropped out of the topographic mapping field. However, a number of Geological sheets produced before 1936 were still available. The 1939 count of both types of topographic maps by province is as follows:[15]

	NTS	Geological
British Columbia	13	10½
Alberta	5	9
Saskatchewan	0	0
Manitoba	3	0
Ontario	86	8½
Quebec	50	5
New Brunswick	4	9
Nova Scotia	24	2
Prince Edward Island	2	0
	187	44
Grand Total 231		

At the end of the series, in 1952 when the last one-mile sheet was published, there were 665 sheets published (15 minute by 30 minute quadrangles). These were distributed over the face of Canada as shown on Figure 14a. A portion of a typical sheet is given in Figure 8.

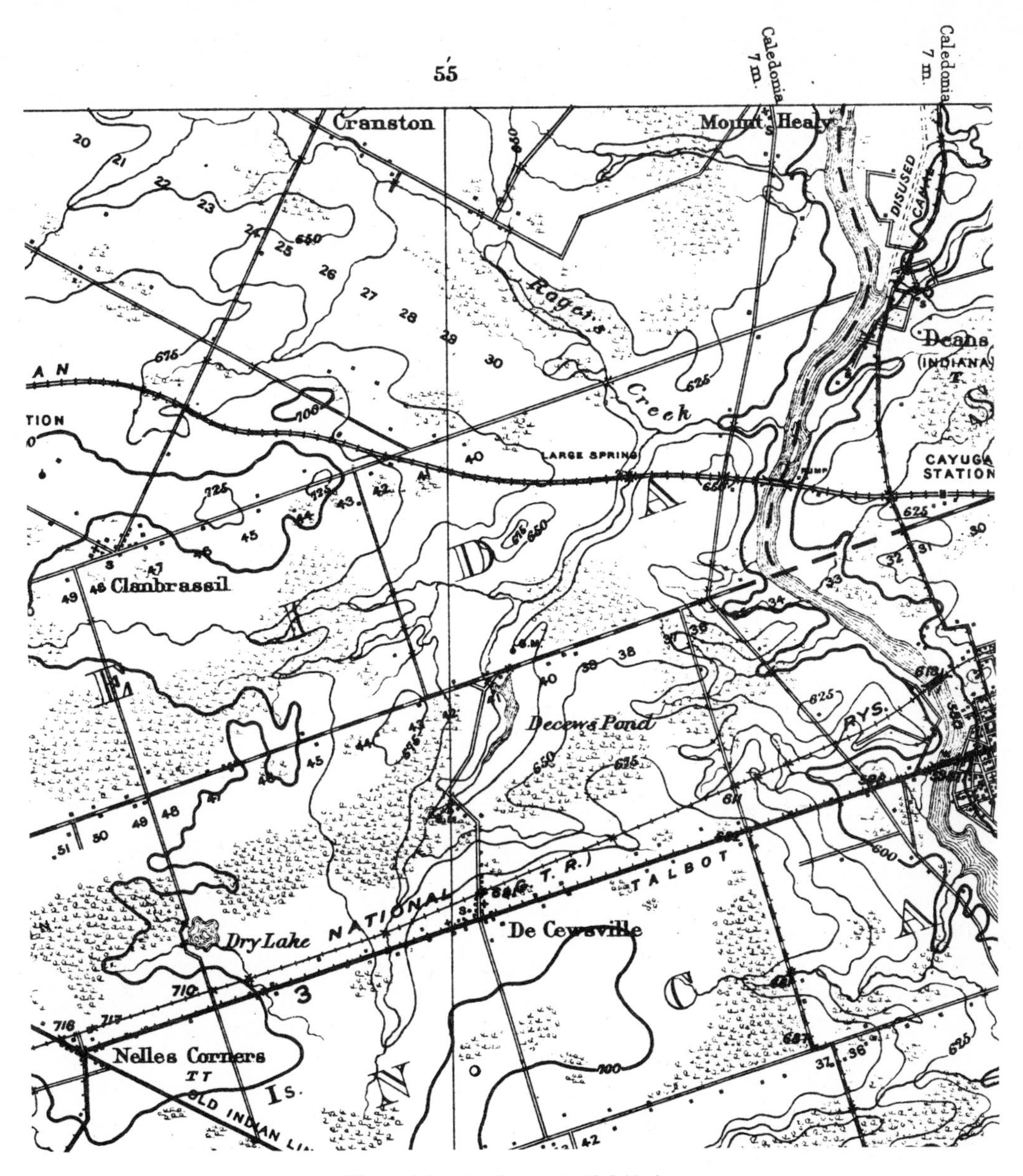

Figure 8 Section from a 1:63,360 sheet

5. The 1:50,000 Series

The 1:50,000 series is Canada's basic topographic map. It is used for planning the development of the northern part of the country and for the administration of the southern part. It is the basic map for national defence, for economic studies and for the protection of the ecology. It has wide use in education and outdoor recreation.

When completed, the 1:50,000 series will have 13,150 sheets of which over 8000 will be published by 1980. As has been mentioned, work on the series was started in 1904 under the one inch to one mile mapping programme. In the intervening 75 years the progress of mapping has advanced from the original five sheets per year produced by Canada's military mapping agency to the present annual production of about 350 new sheets and 300 revised sheets. The change from the quiet production of a few elegant maps modelled on the Ordnance Survey maps of England to the present production covering all types of terrain, wheat fields, boreal forest, glaciers, metropolitan areas, has not been without difficulty. Some critics are scornful of the numerous changes in style that took place while the mapping authorities struggled to adapt the series to the demands of a country that suddenly began taking stock of its wilderness heritage. Other critics point to the lack of definition in the drawing of such features as the forests, agricultural land and the great swamps. The maps of small countries are held up as excellent examples (Finland defines six types of forest cover, Sweden distinguishes cultivated land, meadow and rough pasture) without considering the manpower that would be required to gather this data for Canadian maps.

In the description of this series, which follows, the development of the present styles will be outlined together with the reasons for the important changes of style that were made along the way.

The History of the 1:50,000 Series

In 1950 it was decided to convert Canada's one-inch to one-mile map series to the slightly larger scale of 1:50,000. The decision was made principally because of a military agreement that Canada had entered into with Great Britain and the United States to standardize military mapping at the metric scales of 1:25,000, 1:50,000 and 1:250,000. Canada was thus faced with the choice of maintaining two completely separate series of maps, one military, the other civilian, or coming to an agreement on a single series that would serve all users. After a period of study it was agreed that the metric scales would be adopted but military and civilian versions of each sheet would be published. Essentially the difference between the two was that the military version would have a 1000-metre grid which would not be printed on the civilian version.

The benefits of the move to the 1:50,000 scale were latent but rather long in appearing. Now that Canada has converted to metric units the advantages of having a scale that can be expressed in the simple relationship of two centimetres to one kilometre are apparent to everyone. But in the early years of the series there was bitter denunciation from almost all classes of users. The awkwardness of trying to visualize 1¼ inches to the mile, and the difficulty in taking off mileage measurements with any household ruler be it in inches or centimetres, were pointed out by many. Some went so far as to see in the move an abandonment of British heritage.

But the greatest disadvantage in the adoption of the 1:50,000 scale was one raised by very few users. This was the unfortunate fact that the enlarging of the standard 1:63,360 sheet to 1:50,000 necessitated a paper size that was too large for the military presses of the day. To accommodate a second military agreement (on map size) all sheets south of the 61st parallel of latitude would have to be published in two halves. (North of 61° the convergence of meridians was sufficient to allow the 30 minute-wide maps to be printed on the military presses.)[1]

The question of half-sheets caused considerable confusion. Each half-sheet was a map in itself complete with all the standard marginal information but both halves had the same sheet name, and only differed by the addition of the words 'East' or 'West' on the serial number. Clients ordering maps would often forget to specify which half they wanted, and more often than not would end up with the wrong map. The small half-sheet format was also unpopular and many users looked upon the change as a waste of paper and printing facilities. Eventually the military restrictions on paper size were relaxed and in 1967 the publishing of the 15 minute by 30 minute quadrangle on single sheets was commenced. From that year on, all quadrangles published in half-sheets due for revision were re-issued as single sheets.

Many agencies that used the federal one-inch series as a base for their own mapping refused for many years to follow the federal lead into metric scales. The Geological Survey of Canada, until 1965, reduced all 1:50,000 reproduction material to 1:63,360. Most provincial agencies did the same, except in Quebec where the metric scales were popular from the start. Today almost all one-inch mapping in the provinces has been dropped in favour of 1:50,000 but because of the long useful life of some geological maps the one-inch geological series will be available for many years to come.

When the decision to adopt the 1:50,000 scale was made, a programme was started to convert the existing 1:63,360 sheets to 1:50,000. The opportunity was taken at the time to do a certain amount of revision of planimetric detail. In most cases this was done simply by scaling off new detail from newly flown aerial photography and plotting the work by visually comparing the new detail with the planimetry of the older sheets. This type of revision does not conform to the rules of strict photogrammetric accuracy, but quite useful sheets were turned out. The enlargement of the reproduction material by photography produced rather coarse line-work which is noticeable on the more heavily contoured sheets. The majority of these have been re-scribed in the intervening years.

For the first fourteen years of 1:50,000 production, two versions of each sheet were printed, a military version with a 1000-metre grid and a civilian version without a grid. This necessitated the warehousing of both products, which was somewhat wasteful in space, but it did permit records to be kept of the usage of both versions. By the 1960s it was found that the convenience of the point referencing system provided by the grid was attracting an increasing number of civilian users to the gridded version. In 1964, it was decided that only the gridded version would be published. As with the previous standardization decisions, this one brought protests from certain segments of the user population. Many objected on aesthetic grounds, but others had more practical reasons. On the Prairies many users pointed out that there already was a point referencing system in the Dominion Land Survey grid pattern (see Chapter 3). To these people the additional grid lines were only a confusion. Others stated that although the grid was printed in light blue, when a portion of a map was photocopied, the grid lines appeared the same as the black lot lines. Nevertheless the grid lines have remained and today very little objection is heard.

Considering that the 1:50,000 series has been in existence for less than 30 years, it has gone through a remarkable number of format and style changes. Table 6 lists the various versions and their principal characteristics. A brief description of each follows using the style name found in the government map catalogues.[2] (Where a blank occurs in the style column of the catalogue, the standard style for the year of publication is to be presumed.)

1950 Standard

This was the format developed by the Canadian Army for the first issue of the 1:50,000 series. Sheets were plotted on 15 minute by 15 minute quadrangles, each of which was designated as an East or West half. The simplified border arrangement gives a clean, almost austere appearance. A substantial legend (in English only) is carried in the right margin. Originally the neat-line served as a border, with 5-minute latitude and longitude ticks providing the geographical reference. A later version uses a double-line spaced at 1.5 mm with alternate black and white minute dashes. The colours and symbols used in this version are those of the 1:63,360 series that preceded it. All sheets were originally published with and without the grid, though on reprints after 1964 only the gridded version was provided. The legend used with this style is almost identical with the legend on the present coloured 1:50,000 sheets shown in Figure 13.

1953 Standard

In 1953 at the urging of the Topographical Survey (the civilian mapping agency) a more formal presentation was adopted.[3] The five-line border and other marginal detail formerly used on the one-inch series, were again employed. The legend was moved to the bottom margin, and in general, other than the size of the sheet, the appearance of the pre-war series was reinstated. This style was continued until 1968, but during this period there were certain changes in symbolization. The differentiation between deciduous and coniferous forest cover (previously shown by symbols) was dropped in 1960 and from that year the

Table 6 1:50,000 Series Style Changes

Serial	Style	Number of colours	Year first published	Year publication stopped	Size of sheets (in minutes)	Number published	Number in use Feb. 1978	Publishing agency
1	1950 Standard	5	1950	New maps 1954 Revisions 1964	15 × 15	653	86	ASE and Topo Survey
2	1953 Standard	5	1953	1968	15 × 15	3900	1457	ASE and Topo Survey
3	Preliminary	2	1952	1959	15 × 15 and 15 × 30	160	36	ASE and Topo Survey
4	Arctic Provisional	2	1956	1965	15 × 15 and 15 × 30	240	208	ASE only
5	1959 Provisional	2	1959	1967	15 × 15 and 15 × 30	1330	1220	Topo Survey only
6	1966 Provisional	6	1966	1974	15 × 30	1340	1202	Topo Survey only
7	1967 Standard	6	1967	Still published	15 × 30	Increasing	1885	Topo Survey only
8	1972 Monochrome	1	1972	Still published	15 × 30	Increasing	1875	Topo Survey only
9	Photomaps Monochrome	1	1967	Still published	15 × 30	Increasing	509	MCE and Topo Survey
10	Photomaps (coloured line enhancement)	4	1972	Still published	15 × 30	Increasing	140	MCE

ASE—Army Survey Establishment
Topo Survey—Topographical Survey of Canada
MCE—Mapping and Charting Establishment
Department of National Defence

Monochrome—black and grey
2-colour—black and blue
4-colour—black, blue, red, orange
5-colour—black, blue, red, brown, green
6-colour—black, blue, red, brown, green and orange

green tint indicated simply that the area was tree covered. At about the same time the contour numbers were arranged to read up-hill rather than right-side up from the bottom of the sheet.[4] However, the contour plates for many of the older sheets were not redrawn for many years and consequently 'up-side down' contours persist on a few sheets to this day.

Preliminary[5]

For various reasons it was necessary during the early days of 1:50,000 production to publish sheets that were deficient in some respect. This deficiency might be a lack of contours, a failure to show forest cover or some other omission. They were published under the 'preliminary' title to avoid a delay while the additional data was obtained. The colours and symbolization of the data that was shown were the same as those being used for the standard map of the day. Today the photomap has taken on the function of this 'emergency' style.

Arctic Provisional

In 1956 the Canadian Army found it had an urgent requirement for a number of 1:50,000 sheets of areas in the Canadian Arctic. This was a region where few man-made features would appear on the maps and the forest cover was practically non-existent. With these facts in mind, and to speed production, a new style was developed in which all detail including single line streams and lake shores was published in black. The only other colour was light blue which was used for open water. The demands of the single colour for all detail caused a new legend to be drawn up, as shown in Figure 9. A very simple margin layout was used, and in many cases the contour numbering was quite obviously done by hand. Because of the remoteness of areas mapped in this style, many of these sheets are still the only 1:50,000 coverage available.

1959 Provisional

The production advantage of the army's Arctic provisional sheets was not lost on the Topographical Survey and in 1959 they issued a provisional style of their own. Like the army's maps it used two colours, black and blue, and the coverage provided was restricted mostly to unpopulated areas. In presentation the 1959 provisional maps were somewhat more sophisticated in appearance than their military predecessors. Dark blue was used for river and stream drainage while light blue (i.e. screened blue) was used for open water. Black was used for all other detail. Roads were symbolized in four classes, all weather roads were shown by a heavy black line, dry weather by parallel fine lines, cart-tracks by a pecked line and winter roads by a pecked line with a label. Forest cover was depicted by a pre-printed transparent tree pattern which was applied to the manuscript and printed in a screened grey. For the first year, sheets of this style were printed in east and west halves, but in late 1960 there was a return to the full 15 minute by 30 minute quadrangle. By this time (1960), the newer military presses could cope with the larger format and no objection

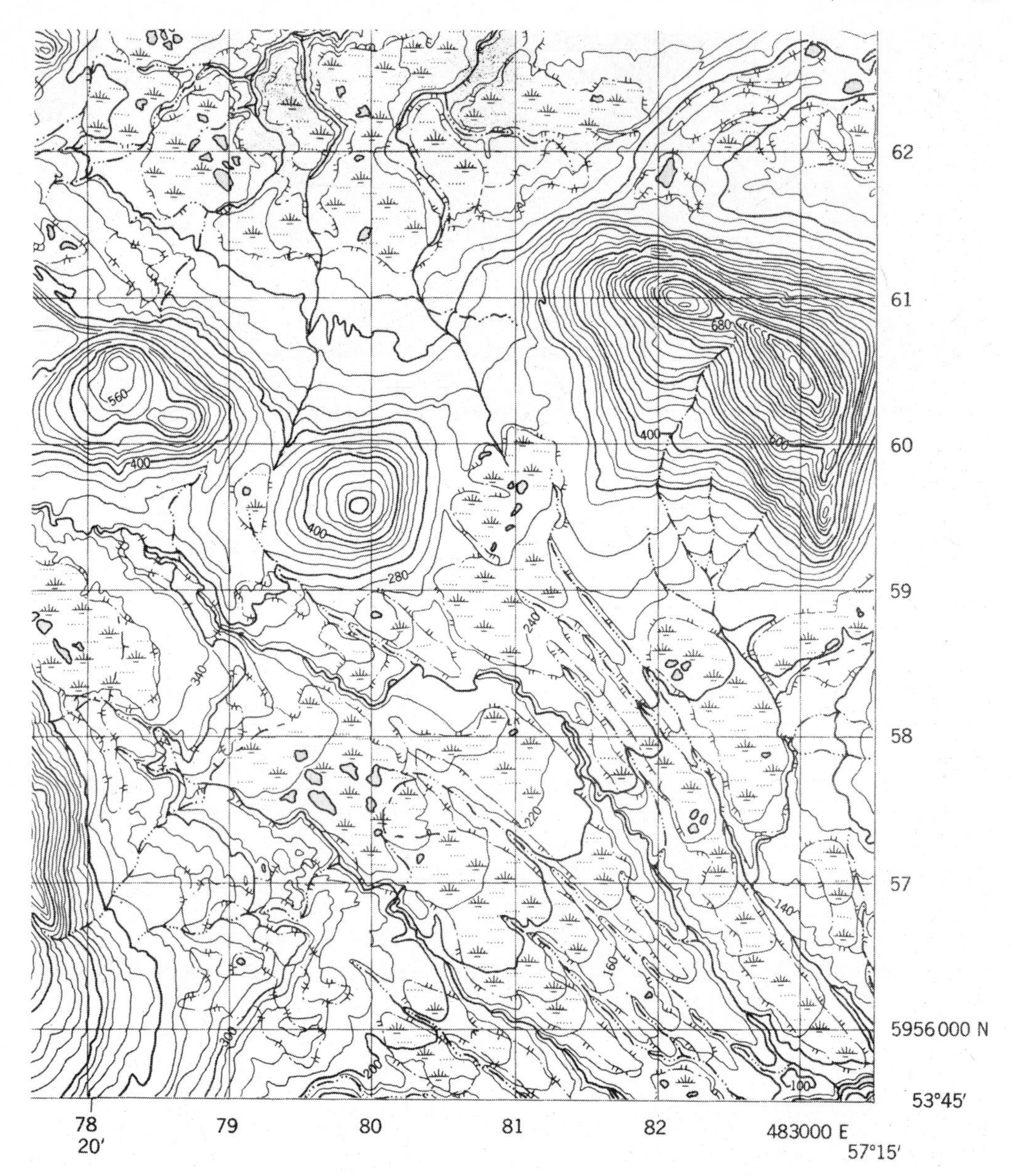

Figure 9 Section from an Arctic provisional sheet

was made to this quiet move by Topographical Survey to go back to the one-inch sheet lines for a 'provisional' map.

1966 Provisional

The 1959 provisional was quite adequate for the portrayal of wilderness areas, but it was not so successful where any amount of cultural detail was to be shown. By 1966 the Surveys and Mapping Branch had acquired sufficient two-colour presses to more than keep abreast of the workload and it was decided to use more colours, but still keep a comparatively simple 'provisional' style. Within the border of the map the data was much the same as on a standard map. The same basic colours were used:

Cultural detail in black
Drainage detail in blue (screened blue for open water)
Contour detail in brown
Forested areas in green

It was in the road depiction that the greatest change took place. Road casings were dropped. Paved roads were depicted by three thicknesses of red line, and gravel and dirt roads were shown by similar lines in orange. This treatment of roads can be inferred from Figure 13 which is a black and white rendition of the legend used with this style. The abandonment of road casings speeded up production of new maps and greatly facilitated revision.

There were other economies made in the drafting required to produce a sheet. The simple single line border was used once again. All names were in black (on standard sheets the drainage names were in blue) and generally, fewer names were employed than would normally be found on a standard sheet. The sheets were, of course, in the 15 minute by 30 minute format. An abbreviated legend showing only the road depiction was used in place of the longer legend on the standard sheet.

1967 Standard

The success of the provisional styles in re-introducing the 30 minute wide quadrangle led to the obvious next step, the 15 minute by 30 minute standard sheet. This was introduced in 1967 with the same colours, symbolization and marginal treatment as the 1953 standard, except that the road depiction that had been developed for the 1966 provisional (red and orange lines without casings) was used. Relatively few of these sheets were produced between 1967 and 1973 because the 1966 provisional style was used for most new sheets. The policy on which style would be used was not well defined, but in general, if the area being mapped had no large towns or cities in it, the provisional style was used. This meant that the standard was reserved mainly for revised editions of older sheets previously published in east and west halves.

By 1972 a serious backlog had developed in the colour separation drafting area of the Surveys and Mapping Branch. This, coupled with the fact that the provisional style being employed was not much more economical than the standard, led to a study of the whole problem of completing the 1:50,000 mapping of Canada. This was an enormous task as the completed series would embrace about 13,150 sheets. The need for six-colour sheets of the far north was seriously questioned. The decision was to drop the 1966 provisional and introduce a monochrome design in its place. The 1967 standard would be retained for sheets of southern Canada.

1972 Monochrome

The area to be mapped in monochrome was formally separated from the areas to be mapped in colour by an irregular line (called the Wilderness Line) drawn across the index

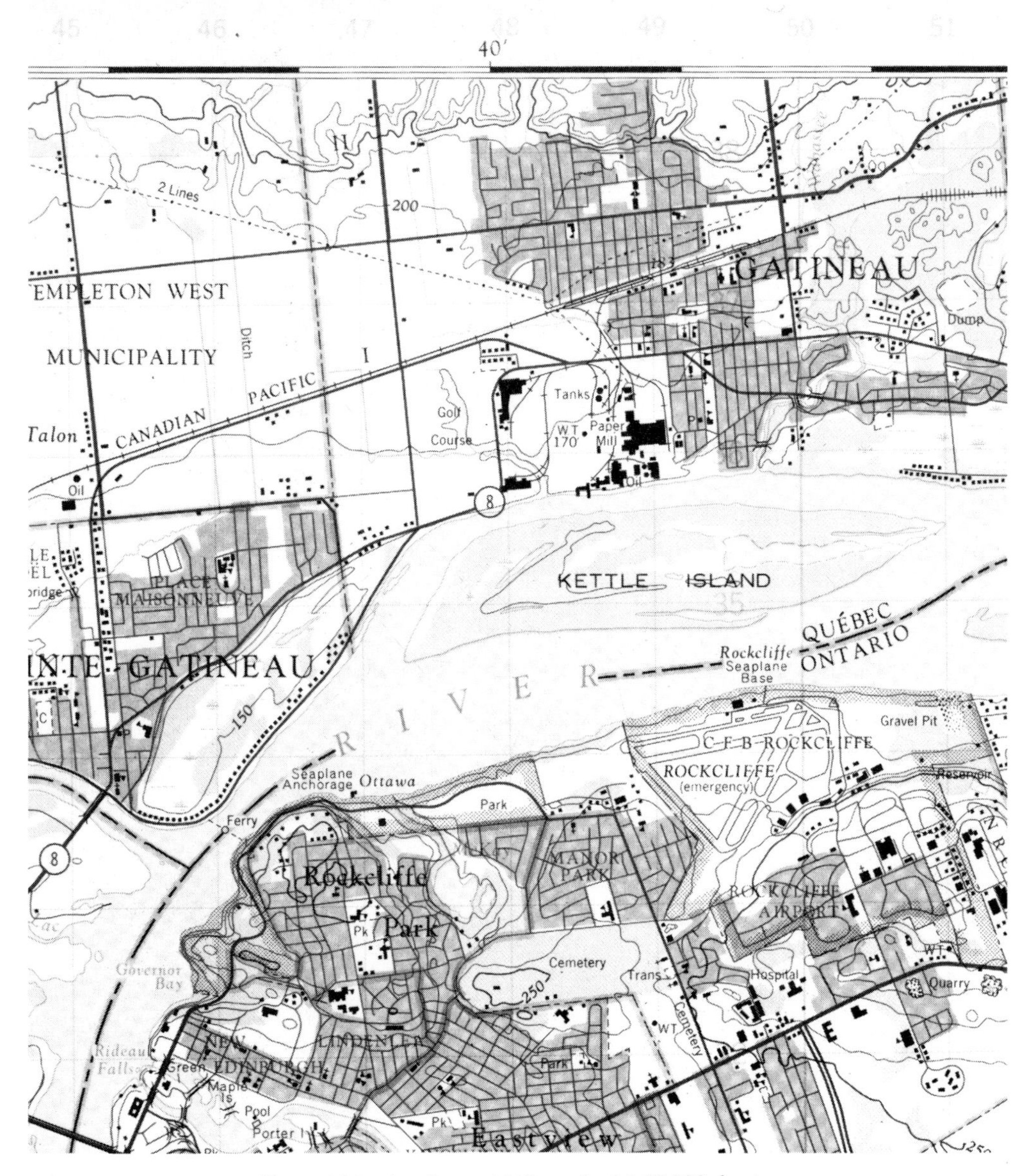

Figure 10 Section from a 1967 standard 1:50,000 sheet

map of Canada, (see Figure 14 A). This line quite arbitrarily separates the generally settled and developed south (areas into which access is reasonably easy) from the undeveloped north where access usually requires the use of aircraft.

Thus we see a repetition of the situation that occurred in the early days of the one-inch series. Two sub-series have emerged and are being produced and maintained side by side. As with the one-inch series, the symbolization of the two sub-series will be examined separately, starting with the more formal coloured maps.

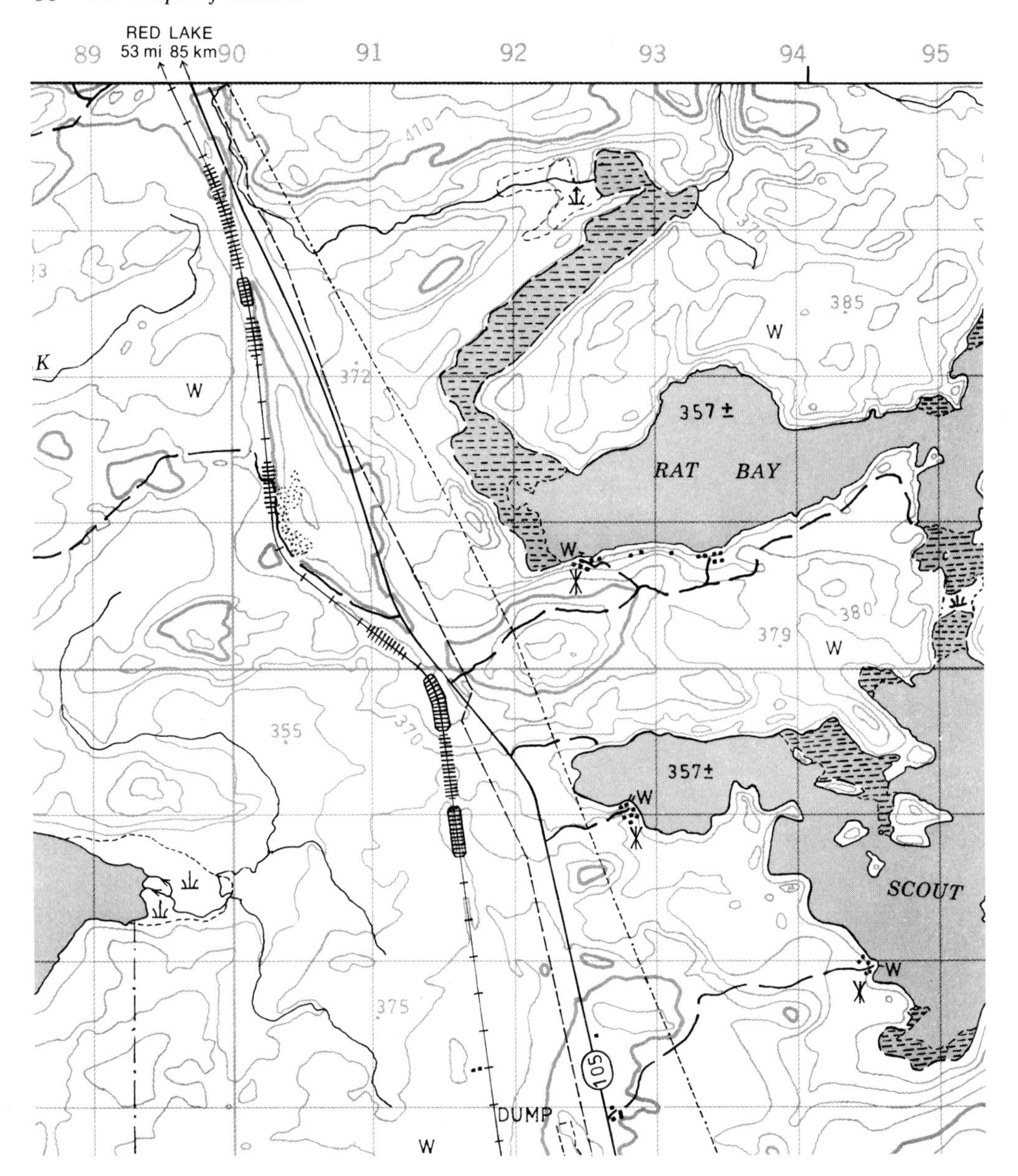

Figure 11 Section from a 1972 monochrome sheet

Symbols and Colours in Use in 1978[6]

Coloured Maps

Figure 10 is a section taken from sheet 31 G/5 Ottawa. The symbols used on the 1:50,000 coloured maps are shown in Figure 13. The following is an analysis of the symbolization and its application to the Canadian landscape.

ROADS. Three colours are used to display the eight classes of roads used in this series.

Red is used for four classes of paved roads, namely:
Dual highways (two parallel 0.4 mm red lines).
Paved roads with more than two lanes (0.75 mm red line).
Paved road with two lanes (0.5 mm red line).
Paved road with less than two lanes (0.25 mm red line).
Orange is used for gravel or dirt roads:
All weather gravel or stabilized road with two or more lanes (0.55 mm orange line).
All weather gravel or stabilized road with less than two lanes (0.35 mm orange line).
Dry weather road (0.35 mm pecked orange line).
Black is used for a cart track or winter road (0.18 mm pecked black line).

This extensive classification reflects the importance of the road network in Canadian life. The movement of people by private car and bus has seriously curtailed the rail passenger service, and except for long-haul bulk cargoes such as ore and grain, most freight is moved on the highways by truck. The winter roads are more frequent in the region of Canada mapped in monochrome, but there are some south of the Wilderness Line. These roads use frozen lakes and marshes wherever possible to reach their destination, but go overland when no water body points in the right direction. Tractor trains are the usual mode of transport, but the better developed winter roads can be used by trucks.

RAILWAYS. Despite the inroads of road transport, the rail network remains a vital feature of the Canadian economy. The depiction of railroads on the 1:50,000 series has not changed from that of the one-inch mapping. Four symbols are used, each based on a black line 0.12 mm in width, depicting single-track lines, multiple-track lines, narrow gauge lines and abandoned lines (as shown in Figure 13). Very few narrow gauge lines remain in service in Canada so that symbol is rarely used. Abandoned lines are shown on new editions only when there is clear evidence on the ground of their previous location. The exception to this rule is in cases where an abandoned railway had an important historic connotation, such as the narrow gauge railway that ran south from Dawson City in the gold-rush days. Stations and flag stops are depicted as black buildings on or across the line, again as shown in Figure 13. Two features of the days of steam trains, the round-house and the turntable, are no longer in use with diesel engines, but the turn-table symbol, curiously, remains in the list of symbols.

HYDROGRAPHY AND WET GROUND. The depiction of river systems is the same as on the one-inch series and in fact is standard for the topographic maps at this scale published by most countries. Rivers are shown as single blue lines if they are narrower than 25 metres and as double blue lines with a light blue open-water fill when they exceed that dimension. Lake shorelines are blue and the water surface light blue. Depths of lakes are not shown but the elevation of the water surface at normal water stage is shown on all lakes over four square kilometres in area. If the elevation has been obtained photogrammetrically it is given in brown, if obtained by field surveys it is in black.

Falls and rapids are symbolized; the former by a bar across the fall-line with wing-marks pointing down stream, the latter by a simple bar across the river. On wide double-line streams, where four or more dark blue ticks can be drawn in the river parallel to the shore, this symbol is used to indicate rapids. On falls, the drop in metres is given when it exceeds three

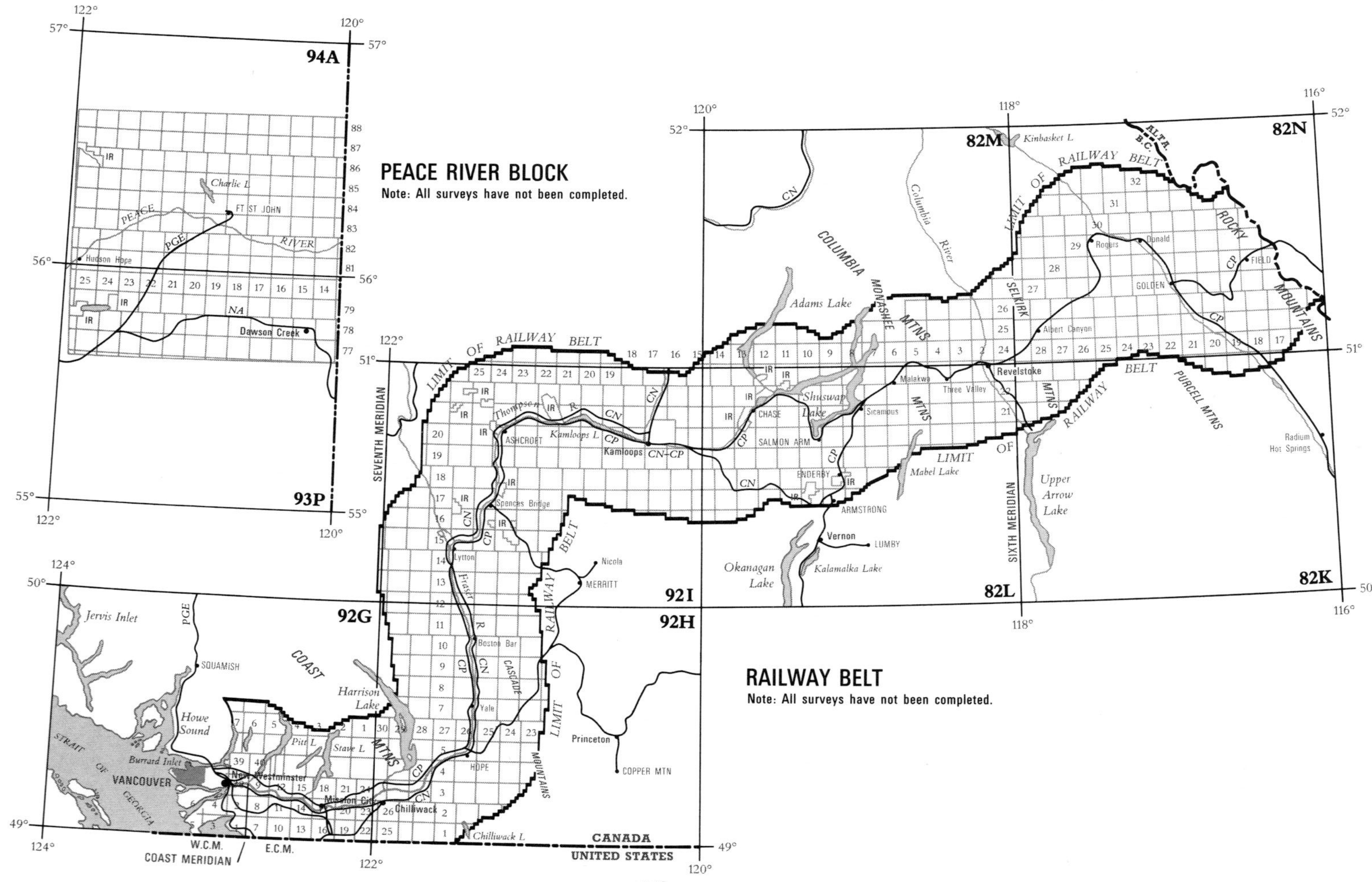

Figure 12 DLS systems—British Columbia

metres. The position of rapids is often accentuated by the addition of the letter R, the abbreviation 'rap' or the word 'rapids' on the shore beside the symbol.

The depiction of wet ground is particularly important on Canadian maps, because Canada has so much of it. Some of the northern swamps are so large that the better known wet grounds of the world, such as the Pripet Marshes, the Sudd or the Okeefinokee Swamp, could be dropped into them without appreciably changing their size. As most of Canada's great swamps lie in the region mapped in monochrome, it might be better to describe wet ground depiction in that section; but, for consistency in presentation it will be included here, and the alternative symbolization required by the single colour will be pointed out when the other monochrome symbols are described.

The Canadian usage for wet ground terms is as follows:

Marsh	Soft wet land commonly covered partially or wholly with water, producing herbage, plants and mosses, but not trees or shrubs.
Swamp	Soft, low ground saturated with water, but not necessarily covered with it, producing trees and shrubs.
Muskeg	A rocky basin filled with successive deposits of unstable materials, such as leaves, muck and moss, often producing swamp spruce (*Picea mariana*).
Slough	A non-boreal lake or pond of shallow depth, liable to dry up at certain seasons of the year.
Bog	A collective term for certain types of wet ground, such as:
Peat bog	Wet, spongy ground containing peat and other organic substances.
Palsa bog	Wet areas filled with hummocks of peat and ice.
String bog	A lake or pond almost completely filled with floating vegetation.

The depiction of these features is as follows:

Marsh	The bulrush symbol interspersed with a horizontal pecked blue line, as shown in Figure 13.
Swamp	The marsh symbol printed over the green forest tint.
Muskeg	This feature is depicted by either the marsh or swamp symbol depending on whether or not trees are present. It is difficult to identify true muskeg (as opposed to swamp or marsh) from the interpretation of aerial photography so the one-inch muskeg symbol was dropped in the 1:50,000 series.
Slough	The intermittent lake symbol is used, see Figure 13.
Bog	The various types of bogs are identified with the symbols shown in Figure 13, but are also usually labelled.

FORESTS AND ORCHARDS. Forested areas, defined as areas where trees are over two metres in height with the crowns covering at least 35% of the ground, are depicted in green tint. No further differentiation into species of growth is made. This has caused insidious comparison with other countries where more elaborate definitions of the forested areas are made. But, with the vast areas of forest to contend with, it was decided in 1958 to limit the forest depiction to the one symbol.[7] Orchards, vineyards and hop fields are the only exception to this solid green symbol. The cultivated vegetation is shown in green as follows:

Orchard	Pattern of 0.4 mm dots.
Vineyard	Coloured screen producing rows of 0.2 mm dots.
Hop field	Same screen as for vineyard, but labelled 'hops'.

Forests are of course very important in the Canadian economy and there has been some criticism of the federal mapping agency for the lack of species identification, even into the fundamental division of coniferous and deciduous classes. As has been mentioned, the drain on available manpower to conduct the necessary forest surveys has been the reason for not maintaining this basic minimum of tree classification. Most of the provinces publish forestry maps (as will be seen in Chapter 13), and these, being drawn purposely to support their forest industry, show the tree stands in much more detail than would be possible on a general-purpose topographic map. One indication that the federal maps need not show more forest information is provided by British Columbia. Because this province, between 1937 and 1977, directly supported the federal 1:50,000 mapping programme by field surveys and map compilation, the provincial mapping agency was given 800 copies of each sheet printed to the provincial specifications. Despite the fact that British Columbia has the largest provincial forest industries in Canada, it requested that the 800 sheets be printed without green so that a special cadastral overprint would show more clearly.

BUILDINGS. Buildings continue to be shown in much the same way as they were on the one-inch maps. Residential buildings and small commercial buildings are shown by a standard black 0.6 mm square, as shown in Figure 13. Large buildings (i.e. those over 200 feet in the major dimension) are shown in plan view to scale. Churches and schools are symbolized, the former as a building symbol with a cross, the latter with a flag. Outside built-up areas, all dwellings are shown, but in a farmyard complex only the farmhouse and the major barns are drawn. Barns on early 1:50,000 sheets were depicted by a small open square, but since 1973 a small solid black rectangle has been used. Factories, shopping centres, paper mills and other landmark structures are labelled. Hotels and motels are labelled but only where they exist outside major towns and cities.

CITIES AND TOWNS. Canadian urban areas normally consist of a heavily built-up core surrounded by less completely developed areas and, further out, ribbon development along the roads leading into the area. On the outskirts, individual buildings are symbolized but when the town or city blocks are more than 60% built-up a screened red tint is used. Landmark buildings of conspicuous size and appearance are still to be shown in the tinted areas.

Boundaries and Cadastral Information

The control and administration of provincial land is in the hands of the individual provinces. This means that each province has its own peculiarities in the subdivision and recording of land holdings.

The 1:50,000 maps show an outline of the land subdivision in as much detail as is practicable for the region on the map. This results in the appearance of Roman and Arabic numerals on the map that are unintelligible without a guide. Table 7 is an outline of the overall situation which will now be described, province by province.

Table 7 Provincial Administrative Boundary Hierarchy

Province	1st administration subdivision	2nd administration subdivision	Sections or farm lots	Special boundaries	Remarks
Newfoundland	None	None	Not shown	Indian and military reserves, parks, etc.	Federal electoral ridings were once used as districts. This was discontinued in 1975 at the request of the province
Nova Scotia	County or District	None	Not shown or numbered	Indian and military reserves, parks, etc.	Nova Scotia has no townships or parishes
New Brunswick	County	Parish	Not shown or numbered	Indian and military reserves, parks, etc.	
PEI	County	Lot	Not shown or numbered	Indian and military reserves, parks, etc.	The PEI lot is of the same size as, and has the function of, a township
Quebec	County or Municipality	Township (canton) or Parish (paroisse)	Rangs and lots numbered in outline	Indian and military reserves, parks, etc.	Rangs are numbered in Roman and lot numbers appear in sufficient frequency so that the sequence may be inferred. Lot boundaries are not shown unless they coincide with a road
Ontario	County, District, Regional Municipality	Township, Township Municipality, Borough, Improvement District	Concessions and lots numbered in outline	Indian and military reserves parks, etc.	Concessions are numbered in Roman. Lot numbers in Arabic appear in sufficient frequency so that the sequence may be inferred. Lot boundaries are not shown unless they coincide with a road
Manitoba	Rural Municipality	DLS township	Section lines shown	Indian and military reserves, parks, etc.	Township and Range numbers appear in margin. In old parishes and settlements that predated DLS Survey an indication of the lot numbering is given
Saskatchewan	Rural Municipality	DLS township	Section lines shown	Indian and military reserves, parks, etc.	Township and range numbers appear in margin. In old parishes and settlements that predated DLS Survey an indication of the lot numbering is given

Table 7—*continued*

Province	1st administration subdivision	2nd administration subdivision	Sections or farm lots	Special boundaries	Remarks
Alberta	County Municipal District	DLS township	Section lines shown	Indian and military reserves, parks, etc.	Township and Range numbers appear in margin. In old parishes and settlements that predated DLS Survey an indication of the lot numbering is given
British Columbia	Land District	Township or District Municipality	DLS system, BC township system and District lots	Indian and military reserves, parks, etc.	DLS system in Railway Belt and Peace River Block. In remote area lots are outlined and numbered
Yukon and Northwest Territories	District	None	Not shown	Indian and military reserves, parks, etc.	Townsites and settlement boundaries are the only boundaries shown within districts

NEWFOUNDLAND. This province is unique in Canada in having no county or district structure. For some years rural federal electoral ridings were used as *ad hoc* districts for the administration of schools, road improvement, etc., and were shown on topographic maps as districts. At the request of the province, this has been discontinued. The title to most parcels of land outside the boundaries of settlements has been obtained by occupation. This has given rise to an infinite number of lot shapes and sizes. As the outlines of lots are uncertain in many cases, no attempt is made to show them.

NOVA SCOTIA. This province has a county structure but no townships. Here again the lots are too small and irregular to be depicted on a general-purpose topographic map.

NEW BRUNSWICK. Counties and townships are found here but once again the lot pattern within the townships is too irregular for depiction.

PRINCE EDWARD ISLAND. In this province the original plantations were so large that they all have been subdivided and now resemble townships. As they were originally called lots, we have the anomalous situation of a township being called a lot and having farms and other holdings within it. The lots are numbered on the map but no further numbering is given to the parcels within the lot.

QUEBEC. In Quebec the land was originally divided into seigneuries. These varied greatly in size but the larger were much the same as the maritime townships (i.e. about 36 to 40 square miles). Many of them were surveyed into rows of farm lots in a manner that was later used in Ontario to provide the township concession system.[8] When the British took over the administration of Quebec, the seigneurial system was allowed to continue until 1847, but even before this date rural land was being divided into townships (sometimes called parishes) for administrative purposes. These townships were then grouped into counties to provide local autonomy in the administration of certain regional functions. In recent years the townships have been found to be too small for efficient administration and many of them have been grouped into municipalities. All administrative divisions are shown on the maps. Within the townships or parishes the lines of lots, called rangs, are numbered with Roman numerals. The lots are not all numbered but Arabic numbers occur in sufficient frequency so that the sequence of lot numbering may be inferred.

ONTARIO. This province was first opened to settlement in 1780 to provide land for Loyalist refugees during the American Revolution. In almost all cases the land was surveyed before it was occupied, and thus a formal township system was set up with regular rows of farm lots that were similar to the rangs of the seigneuries. These concessions (as they are called in Ontario) are numbered with Roman numerals. As in Quebec, the lot numbers appear in sufficient frequency to show the sequence of lot numbering. In the northwest part of Ontario, some townships were laid out by Dominion land surveyors in the manner used for surveys on the Prairies (see Chapter 3). This occurred because of uncertainty as to the position of the western boundary of the province. Later, when the boundary had been established, Ontario land surveyors made additional township surveys using a pattern similar to that of the DLS but using the concession and lot system to designate farm lots.[9]

Originally, in the settled parts of the province townships were grouped into counties, while in remote areas townships were included in districts. In recent years the Ontario government has found that this simple two-tier municipal structure was inefficient, particularly in heavily populated areas. New administrative groupings were made, which at times conflict with the historic boundaries and tend to cause confusion unless the terms used are clearly understood. The following is a short glossary:

First administrative level

County	This is the original southern Ontario first-tier administrative area. It is retained in its original form except in regions of high population density. It may contain cities, towns, townships and villages.
District	This is the original northern Ontario first-tier administrative area, and in general the districts have not been changed in recent years. They may be partly or completely subdivided into townships, and may contain cities, towns and villages.
Regional municipality	This is the new administrative area that replaces the county in heavily populated areas. They may contain cities, towns, township municipalities, townships and villages. All are in southern Ontario except Sudbury Regional Municipality which is entirely within Sudbury District.
Municipality of Metropolitan Toronto	This is the special municipality containing Toronto and its suburbs. It contains the City of Toronto together with its boroughs, each having the status of a township.
District municipality of Muskoka	This is for all intents and purposes a regional municipality. Its title stems from the fact that it was a district (rather than a county) before it was converted. It contains towns and townships. It is a heavily populated summer resort area directly north of Toronto.

Second administrative level

Township municipality	This is one of the normal subdivisions of a regional municipality (along with cities, towns, etc.). Generally, the township municipality has absorbed two or more of the original townships. As the latter must be retained for record purposes (the whole land survey fabric is based on them) they are marked on maps and labelled as 'geographic townships' to distinguish them from the township municipalities.
Township	This term is used to describe one of the original townships that has not been absorbed into a township municipality.
Improvement district	Within the districts of northern Ontario a few improvement districts have been created to provide administrative control in areas where there has been a sudden growth due to the opening of a mine, paper plant or other resource industry. Improvement districts provide second-level administration in such areas until the residents want to establish a village, town or township.

On the topographic maps of Ontario these administrative units are carefully identified by name and designation. At times the following abbreviations are used:

Township	Tp
Geographic Township	Geog Tp
Township Municipality	Tp Mun
District Municipality	Dist Mun
Regional Municipality	Reg Mun
District	Dist
County	Co

THE PRAIRIE PROVINCES. The origin and function of the DLS system has been described in Chapter 3. The township and section system is instantly recognizable on 1:50,000 maps. The township and range numbers are given in the margin while the control meridians and base lines are labelled on the map. As has happened in Ontario and Quebec, the townships are being grouped into larger units (rural municipalities or districts) by the Prairie provinces. These are indicated by the county line symbol and are labelled.

BRITISH COLUMBIA. This province is unique in several ways. First, it has two areas where the DLS system is used. These are the Railway Belt and the Peace River Block shown in Figure 12. In these areas the normal DLS designations are used as described in Chapter 3.

The remainder of the province is divided into land districts which are similar in function to the counties of Quebec or Ontario. Some of the land districts are surveyed, at least partly, into townships that bear a close resemblance to the DLS system, but this is because both were copied to a large extent from the American system of public land survey. It can be said that the British Columbia (BC) imitation came first because township surveys were commenced in 1859 (12 years before the federal surveyors started township surveys on the Prairies), but in the BC case the townships were originally an imitation in miniature as the township was only three miles square, and each section only 160 acres. It was not until 1873 that the six-mile square township was introduced into British Columbia. The BC townships, both three-mile square and six-mile square, are numbered east or west of the

Dual highway
Road, hard surface, all weather, more than 2 lanes
Road, hard surface, all weather, 2 lanes
Road, hard surface, all weather, less than 2 lanes
Road, loose or stabilized surface, all weather, 2 lanes or more
Road, loose or stabilized surface, all weather, less than 2 lanes
Road, loose surface, dry weather or unclassified streets
Cart track or winter road
Trail, cut line, portage — por
Railway, single track
Railway, multiple track
Railway, narrow gauge
Railway, abandoned
Railway station; stop; turntable
Bridge; footbridge
Bridge: swing, draw, lift
Tunnel
Ferry
Ford
Navigation light
Seaplane base; seaplane anchorage
Vessel anchorage: large, small
House; barn; large building
Church; school
Post office; telegraph office — P. T.
Elevator; greenhouse — E.
Cemetery; historic site; historic battlefield — C
Tower, chimney, similar objects
Windmill or windpump
Pipeline: above ground, underground
Telephone line
Power transmission line
Campsite; picnic site
Retaining wall: large, small
Mine
Gravel or sand pit; quarry
Dyke; fence
Cutting; embankment

International boundary with monument
Province, territory, or state boundary
County or district boundary
Township, parish, borough boundary
Township boundary, unsurveyed
Metropolitan area boundary
City, municipality or parish (Quebec) boundary
Reserve, park, etc. boundary
Section line
Surveyed line, lot line
Horizontal control point
Bench mark with elevation — BM 1475
Spot elevation: precise, non-precise — •2520 •2247±
Stream or shoreline, indefinite
Irrigation canal, drain, ditch
Direction of flow
Lake intermittent, slough
Flooded land, seasonally inundated land
Marsh or swamp
Dry river bed with channels
String bog; palsa bog
Tundra ponds; tundra polygons
Falls
Rapids
Reservoir, dugout, swimming pool
Foreshore flats
Submerged reef
Rocks
Rocky ledge, rocky reef
Dams: small, large
Locks: small, large
Wharf, breakwater, pier, dock; seawall
Ferry slip; dry dock; ramp
Glacial crevasse
Wooded area
Contours
Approximate contours
Depression contours
Cliff
Esker
Sand; sand dunes
Moraine, scree
Distorted surface

Figure 13 Symbols for current coloured 1:50,000 sheets

coast meridian which runs north from the point where the 49th Parallel meets the Pacific Ocean at Semiahmoo Bay.

It must be understood that only a small portion of British Columbia is divided into townships. The more heavily populated tracts that lie outside the townships are laid out in district municipalities which have the administrative facilities of the townships without the rigid survey pattern. In more remote areas of the province where there is a need to allocate land for sale, lease or license, 'district lots' are surveyed. At one time the boundaries of these independent lots were laid out on specific bearings and were of specified dimensions, but gradually these restrictions have been abandoned. The survey of BC townships and lots have been sorted into nine different systems of survey. For those interested in the details of this work, the best text is that written by W. A. Taylor in 1975 while he was Surveyor-General of the province.[10]

The 1:50,000 sheets of British Columbia show far more cadastral information than those of the other provinces. This stems from an arrangement set up in 1937 in which it was agreed that the province would assist in the surveying and compilation of certain sheets in the one-inch series.[11] In return the province had some say in the specifications for BC sheets, and opted for more cadastral information, specifically the showing of lot lines and lot numbers.

The lot numbering system used by British Columbia gives a valuable insight into the land use of the area, because many of the lot numbers incorporate a letter code which indicates the intended use of the land under license. For example, CL507 means that lot 507 is registered as a coal license; TB376 is a timber berth. These abbreviations are identified in the marginal information of the provincial maps (see Chapter 13), but unfortunately no explanation is given on the federal 1:50,000 series. The abbreviations currently being used (1978) are as follows:

BK	Block
CG	Crown Grant
CL	Coal License
CM	Coast Meridian
CTP	Christmas Tree Permit
FUP	Free Use Permit
IR	Indian Reserve
L	Lot
LRO	Land Registry Office
MC	Mineral Claim
PL	Pulp Lease
PML	Placer-Mining Lease
PHU	Pulp Harvesting Unit
R	Range
RS	Ranger Station
Res.	Reserve
S or SEC	Section
SL	Sub Lot
SP	Sample Plot
STL	Special Timber License
TB	Timber Berth
TFL	Tree Farm License
TL	Timber License
Tp	Township
X or A	Timber Sale

In addition, certain abbreviations are used with the serial number of survey monuments in wilderness areas to assist prospectors and resource engineers in finding and identifying the monuments. The most common of these are:

IP	Iron Post
AP	Aluminum Post
AD	Aluminum Disc
M	Mound (a mound of earth or rock, always used in addition to one of the other descriptions, such as in IP304M)

YUKON AND NORTHWEST TERRITORIES. The DLS system was extended into the Northwest Territories, but in 1975 it was decided to cease using these few surveys as the basis for legal descriptions of land parcels. Since that time reference to the DLS system has been expunged from 1:50,000 maps. The Territories use a 1:30,000 series (either photomaps or enlarged 1:50,000 maps) for land administration.

A General Comment on Cadastral Information

The depiction of the cadastral system of the provinces and the Territories is not intended to be complete in the 1:50,000 series. Each province has its own cadastral map series, and on these sheets the parcels that can be conveniently shown to scale are identified. But the system of land administration at all levels is most important to those living and working in rural and wilderness areas. The topographic maps would be much less useful if they did not show the cadastral pattern, at least in its outline form.

Marginal Information

The 1978 standard 1:50,000 carries more information in its margin and on the back of the map than any previous style of Canadian topographic map. The top margin of the map consists of the series title, with 'Canada' at the centre of the sheet; the sheet number is at both the north-east and north-west corners, and the scale (1:50,000) and edition number in the west and east mid-point of the top margin respectively. The west margin is blank. The east margin starts at the top with the military reference box which identifies the series, map and edition according to the NATO code. Beneath this is an English–French glossary of about 20 of the most frequently used labels on the map. (A complete glossary of over 300 terms which include virtually all features that are described on a Canadian map is printed in light grey on the back of the map.) Under the glossary is a list of the abbreviations used on the map. This is followed by the magnetic north, true north, grid north diagram and data on the annual change of the magnetic declination. Below this is found instructions on the use of the rectangular grid and a diagram showing the numbers (but not the names) of adjoining sheets. At the bottom of the margin the sheet title is given once again, together with the sheet and edition number.

The bottom margin starts at the left with the English imprint note which contains the name of the producing agency (Surveys and Mapping Branch, Department of Energy, Mines and Resources), the year in which the aerial photography used in the compilation or revision was taken, the year of the culture check, if one was made, and the publication date. Under this is a note that this map can be obtained from the Canada Map Office or from map dealers. Under this is the copyright note.

Moving to the right along the south margin, the next feature encountered is a partial legend consisting only of the conventional signs for roads and tracks. The map user is directed to a complete legend which is printed in grey on the back of the map. At the centre of the south margin one finds the sheet title together with the county or district (if appropriate) and the province or provinces that are shown on the map. The bar scales showing miles, metres and yards are printed below the sheet title.

ROAD, HARD SURFACE, ALL WEATHER
ROAD, LOOSE SURFACE
CART TRACK, WINTER ROAD
TRAIL, CUT LINE, PORTAGE
BUILT-UP AREA
RAILWAY: SIDING, STATION, STOP
BRIDGE
SEAPLANE BASE; SEAPLANE ANCHORAGE
HOUSE; BARN
CHURCH; SCHOOL
POST OFFICE
TOWER: FIRE, RADIO
WELL: OIL, GAS
TANK: OIL, GASOLINE, WATER
POWER TRANSMISSION LINE
MINE
CUTTING; EMBANKMENT
GRAVEL PIT
INTERNATIONAL, PROVINCIAL BOUNDARY WITH MONUMENT
COUNTY, DISTRICT BOUNDARY
TOWNSHIP, PARISH BOUNDARY
D.L.S. TOWNSHIP CORNER: SURVEYED, UNSURVEYED
D.L.S. SECTION CORNERS
MUNICIPALITY BOUNDARY
RESERVE, PARK, ETC. BOUNDARY
HORIZONTAL CONTROL POINT

BENCH MARK WITH ELEVATION
SPOT ELEVATION, PRECISE: LAND, WATER
STREAM OR SHORELINE; INDEFINITE
DIRECTION OF FLOW
LAKE; INTERMITTENT LAKE; PONDS
FLOODED LAND
MARSH; SWAMP (WOODED)
DRY RIVER BED WITH CHANNELS
STRING BOG
TUNDRA: LAKES IN TUNDRA; POLYGONS
RAPIDS; FALLS; RAPIDS
FORESHORE FLATS, SAND IN WATER
ROCKS
DAM
WHARF
DITCH
CONTOURS
APPROXIMATE CONTOURS
DEPRESSION CONTOUR
CLIFF
SPOT ELEVATION, APPROXIMATE: LAND, WATER
ESKER
SAND, SAND DUNES
PALSA BOG
WOODED AREA, FOREST
CLEARED AREA

Figure 14 Symbols for current 1:50,000 monochrome sheets

Moving again to the right, the next feature encountered is the technical data note which gives the contour interval, the vertical and horizontal datum and the projection used for the map. A metres to feet conversion scale and a note telling users that precise elevations of bench marks are available from the Geodetic Survey are also included. The imprint note is repeated in French in the right edge of the south margin.

The border of the map is almost identical with that used on the military one-inch map as described on page 44. Therefore on the 1978 sheet we find, reading out from the neat line, the latitude and longitude bars consisting of two parallel black lines about one millimetre apart which enclose alternating black and white minute bars. About one centimetre outside this we find a fine black line and a thick (1mm) black line which completes the border. In the one centimetre space within the border, we find the grid numbers, the latitude and longitude values and the destination notes.

Monochrome Maps

When the monochrome sub-series was designed in 1972 the attempt was made to keep, as much as possible, the same symbolization as that used for the 1:50,000 coloured sheets. Figure 14 shows the monochrome legend and it will be noted that black is used for both the black and dark blue linework of the coloured sheets, while grey lines replace the brown contours. Black symbolization alone is used for the road classification, replacing the red and orange linework. As Figure 14 portrays the monochrome legend exactly as it appears on the maps, the abbreviated description which follows will suffice. Figure 11 is a section from a monochrome sheet, 52 K/6 Wabaskang Lake.

ROADS AND RELATED FEATURES. Roads are few in the regions mapped in monochrome and therefore the three categories, hard surface, loose surface and winter road or cart track are adequate to portray the road network. In the unlikely occurrence of a divided highway appearing on a monochrome sheet, it can easily be depicted by parallel black lines. Winter roads are common in this part of Canada as they form an economical means of transporting heavy and bulky cargoes once the rivers and lakes are sufficiently frozen. The trail symbol (Figure 14) is used for trails, portages, cut-lines such as fire guards around settlements, seismic lines put in by bulldozer during geophysical exploration, and in fact for depicting any occurrence where a line or trail is cut through the bush. Railroads, being drawn in black on the coloured maps, are shown with exactly the same symbols on the monochrome.

Telephone lines are not common north of the Wilderness Line, and where they do exist they are along roads and hence are not shown. Microwave transmission is the normal method of transmitting telephone messages in the north, and the towers required are depicted by the tower symbol with a label. Telegraph lines have virtually ceased to exist anywhere in Canada. Telegraphic messages are normally passed on telephone lines.

Power lines are important in the north. They are an essential part of most resource development, and are often indicators of the direction of future settlement. From a cartographic point of view, they are clear and unambiguous landmarks to such an extent that a new trunk line crossing a published sheet is usually sufficient reason for an overprint revision. (Overprint revisions will be described near the end of this chapter.) The symbol for a power line, shown in Figure 14, is the same as on a coloured map.

HYDROGRAPHY. All water features are shown in black or grey; black is used to replace the dark blue of single-line streams and lake shores, while grey is used in lieu of the light blue open water fill. As has been mentioned, the greatest swamps in the world are found in northern Canada, so large regions are covered with the swamp and marsh symbols (these will be described in the discourse on vegetation which follows). Rivers and streams passing through these vast areas of wet ground often have indistinct water courses, so many must be shown by the indefinite stream symbol (pecked black line).

VEGETATION. The land north of the Wilderness Line is a land of vast forests, muskeg, and swamps until the tree line is reached. North of the tree line lies the tundra, thousands of square miles of treeless plain. In the tundra the sub-soil is permanently frozen, but the topsoil supports mosses, lichens and bushes. The frozen under-layer prevents seepage into the ground and where surface drainage is inadequate, distinctive ground features develop. Examples are, palsa bogs (bog-like areas filled with hummocks of peat and ice) and tundra polygons (areas of clay and vegetation which are divided into polygons by surface drainage).

The depiction of the various types of northern vegetation is important on northern maps because each type presents different problems to the engineer charged with building roads, pipelines, dams and the like in the area. The depiction of vegetation without the green tint is accomplished by using outlines around the various vegetation types. Within the circumscribed area, symbols are used to define the type of growth. The forest outline is a solid black line with two ticks on the wooded side of the line at intervals of two or three centimetres. This 'forest line' is used to depict the edges of both clearings in forested areas and small woods or copses in generally clear areas; the ticks along the line indicate on which side of the line the forest lies. The forest line is not used where the edge of the forest coincides with other planimetric detail such as a road or stream. Interspersed throughout the forested areas are the letters F (forested), while the letter C (clearing) is used for open areas. Before the bilingualization of the monochrome series, the letter W (wooded) was used in place of F. Marshes are outlined by a pecked black line and are identified by the usual bulrush symbol printed in black. For swamps, the centre line of this symbol is replaced with a coniferous tree symbol to indicate forested wet ground (see Figure 14). String bogs are shown by the bulrushes over wavy lines to give the impression of floating vegetation. Palsa bogs have the letters PB within the outline.

RELIEF. Relief is portrayed by contours, spot heights and the special symbols for eskers and pingos. Contours are printed in grey so that there is a clear distinction between them and the other line work such as that for streams and forest edges. Every fifth contour is an index contour and is printed in a heavier line weight. Approximate contours are pecked lines and depression contours have inward ticks in the same fashion as on coloured maps. Eskers are glacial deposits in the shape of a ridge which may vary from a few metres in height to over 25, and in length from a few hundred metres to many kilometres. Where they exist, they are a distinctive landmark and are shown by the chevron symbol illustrated in Figure 14. Pingos are 'ice volcanoes' resulting from ground water being trapped above a permafrost layer and being forced under considerable pressure to the surface where it freezes and builds to a mound which at times exceeds 50 metres in height. The surface of the pingo is covered by layers of dirt in which Arctic plant life takes root giving the mound a permanent appearance and in fact, making it permanent by providing insulation.

Administrative Boundaries and Cadastral Information

Figure 14 illustrates the boundary symbolization and it will be seen that this differs very little from similar symbols on coloured maps.

Cadastral surveys are shown if they exist in the area being mapped, but, in actual fact, few of the Quebec and Ontario surveys extend north of the Wilderness Line. The Dominion Land Survey system of the Prairies and northern British Columbia are, on the other hand, a prominent feature on the monochrome maps, and are prominently and rather strangely portrayed. Township corners are marked with large black circles nine millimetres in diameter. If the corner has been fixed by survey, the circle is solid and a black cross is placed at its centre. If the corner is in its theoretical position, but no survey has reached it, it is shown as a pecked circle with no cross. Section corners along surveyed lines are shown by black crosses. Theoretical township corners are only depicted if the base line to the north of them has been surveyed. (See also Chapter 3.)

Marginal Information

In keeping with the provisional nature of monochrome maps, the marginal information is kept as simple as possible. The neat line serves as the border of the map. The grid numbers lie outside the neat line as do the five-minute values of latitude and longitude. Small 1.2 mm black circles are placed on the neat line at each five-minute mark. A full legend is provided in the right margin. Beneath the legend is a diagram showing the photography used in compiling the map, and if the map is a second edition, the photography used in the revision. It is anticipated that much of the field-use of these sheets will be by people also using aerial photographs, and in such cases these diagrams will be most helpful. The standard instructions for using the grid, a guide to adjoining sheets, a magnetic variation statement and a 'feet-to-metre' conversion scale completes the right margin.

The left margin is clean. The top margin is restricted to the word 'Canada' and the edition and sheet number.

The bottom margin starts at the left with the credit note in English. The sheet title and the province or territory in which the map falls is at the centre. The bar scales in miles, metres and yards are under the title block. The credit note is repeated in French at the lower right corner of the neat line. At the lower right corner of the paper, the sheet number and edition number are repeated for the convenience of persons looking for a particular sheet on shelves or in map drawers.

Revision

The government publication of a topographic map carries with it the implied responsibility of keeping that map up-to-date. As the coverage provided by the 1:50,000 series increases, the revision load increases, but not at a proportional rate. The new maps are entirely of wilderness areas where the rate of change in the topography is slow.

For convenience in 1:50,000 revision planning, the whole of Canada has been divided into areas where the rate of obsolescence of maps is approximately the same. These areas are as follows:

5-Year Cycle Area*	198 sheets covering cities and their suburbs and outlying areas.
10-Year Cycle Area	1450 sheets covering rural Canada including all farm lands, recreational areas such as parks, cottage regions, ski resort areas, etc., and those sheets covering Arctic settlements.
15-Year Cycle Area	The remaining 1562 sheets covering Canada south of the Wilderness Line and an additional 1039 sheets covering northern communication routes.
30-Year Cycle Area	The remaining 8901 sheets covering the region to the north of the Wilderness Line.

In revising 1:50,000 sheets, one of three distinct methods may be employed, all of which use new aerial photography of the area being revised. If the dimensional accuracy of the old sheets is questionable, or if the changes to be made are very extensive, the map is completely recompiled. If the framework of the old map is sound, the required changes on the draftings of the various printing colours are made using non-stereoscopic optical devices which allow the image of the new aerial photos to be 'fitted' to the old manuscript. The second method is sometimes referred to as 'desk top revision'. Finally, if the changes are small, and in general are point or linear features, such as new buildings, pipelines, isolated roads etc., the map may be given an overprint revision. When this is done the new features to be added are scribed on a new piece of reproduction material which provides an

* These areas will change with the spread of the cities and the development of new farming and recreational areas, and new communication routes into the north. The figures given are those in use in 1979.

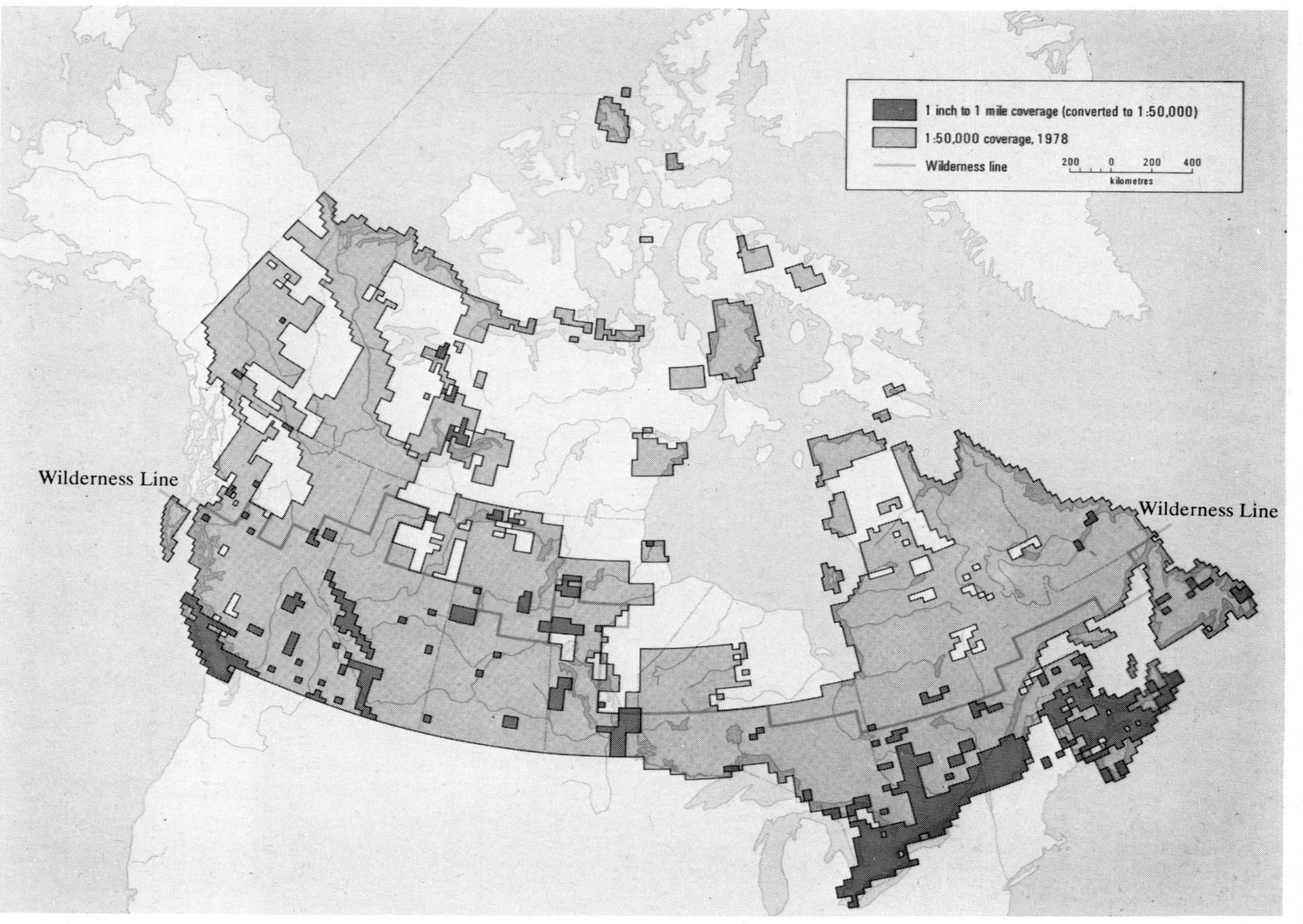

Figure 14A 1:63,360 and 1:50,000 status map

additional colour, purple, for the map. The map is then reprinted with the standard colours remaining the same, or with possibly a few minor deletions, and with the purple colour appearing as an overprint.[12]

In some cases the cyclic revision photography reveals that no changes have occurred in the topography of a particular sheet. In such a case the stock of the sheet is overprinted with a new publication date and the catalogues and indices are amended appropriately.

Photomaps

Photomaps are included in this chapter because the 1:50,000 scale is the only one at which the federal mapping agencies have produced such maps in quantity for sale to the public. Certainly photomaps have been produced at other scales in other times, but these were in answer to requests from special clients, generally other government agencies, and the products were rarely available to the public.*

The 1:50,000 photomap series produced for general distribution was initiated in 1962 by the Surveys and Mapping Branch. The areas chosen for this type of mapping were in the north, in regions of active resource development. As photomaps portray the ground in 'photographic detail' they are useful for land navigation in wilderness country where their minute definition of the ground provides a wealth of landmarks. But they leave a great deal of interpretation of the terrain to the user who must decide for himself whether the ground is dry or swampy, whether the vegetation is brush or forest, etc. Many experienced travellers whose work takes them into the north prefer to use the air photos themselves, rather than the photomap, because with a pocket stereoscope a three-dimensional magnified view of the ground can be obtained. Because of a general lack of demand, the production of 1:50,000 photomaps has been curtailed, though special purpose photomaps are still provided, on request, to government agencies. An example is the series of 1:30,000 whiteprint photomaps provided at the request of the Government of the Northwest Territories for land administration.

The majority of the Surveys and Mapping Branch photomaps are uncontoured. Enhancement is kept to a minimum, and is usually little more than a UTM grid and a few prominent place names. A second and quite different photomap series was produced by the Mapping and Charting Establishment between 1971 and 1976. This series of 140 sheets covers the settlements in the far north listed in Table 8. The sheets are lavishly enhanced with contours in orange, roads and trails in red, open water in blue, and names, spot heights and the UTM grid in black.

The accuracy of the maps of both the civilian and military series varies. If no 1:50,000 line map exists, the photomap was laid down to an enlargement of the 1:250,000 map of the area and therefore would be no more accurate than the source map. If on the other hand 1:50,000 survey data was used, in most cases the photomap would have the same accuracy as the 1:50,000 line map. The military photomaps contain an accuracy statement in the bottom margin.

Progress of the 1:50,000 Series

Figure 14A shows the sheets of the 1:50,000, both coloured and monochrome, that were published up to August 1978. A breakdown of numbers of published sheets, by provinces

* One exception is the 1:25,000 photomap series (29 sheets) of Montreal published in 1965. This was to be the first of an extensive urban photomapping project, but lack of public interest caused curtailment.

Table 8 Photomap Coverage of Arctic Settlements

	Latitude	Longitude	Sheets published
Baker Lake	64°	96°	14
Coral Harbour	64°	83°	14
Frobisher	64°	69°	22
Cape Dorset	64°	76°	3
Pangnirtung	66°	65°	11
Cornwallis	75°	96°	11
Pond Inlet	73°	77°	3
Mould Bay	76°	120°	11
Sachs Harbour	72°	125°	7
Holman Island	71°	118°	5
Isachsen	79°	104°	19
Eureka	80°	86°	14
Tanquary Camp	81°	78°	6
			140

and territories, is given in Appendix 2. By 1980 all the 3200 sheets lying south of the Wilderness Line will have been published. Forecasts of future production of 1:50,000 sheets are difficult due to changes in programmes caused by unexpected government requirements, but the following benchmarks should be achieved:

Completion of the mapping of provincial lands by 1985.
Completion of the mainland of Canada by 1990.
Completion of Canadian territory by 2000.

Over 75 years of hard work have gone into the 1:50,000 series and almost without doubt the year 2000 will see the coverage complete and all editions reasonably up-to-date. There is a mass of information on these maps, and it is certain that they will play an important role in the future growth of Canada.

6. The Two-Mile and 1:125,000 Series

The two-mile series (1:126,720) and its successor the 1:125,000 series are members of an intermediate scale of mapping that have had chequered fortunes in Canada. British Columbia is still producing 1:125,000 and since 1979 1:100,000, topographic maps thus perpetuating a series that was started in that province (as two-mile mapping) in 1912. But the series has been virtually abandoned by the federal government.

Before 1912 a number of two-mile maps were published in Canada, but in general they were drawn up on *ad hoc* specifications to fill an immediate need. There was no long-term plan to cover any amount of territory in an orderly fashion. The scale was popular with geologists and in the early years of this century we find both the Geological Survey of Canada and the Ontario Department of Mines using this scale when the geological formation under investigation was of a size and importance to warrant such mapping. The Québec Départment de la Colonisation produced a number of two-mile maps of the frontier counties, and these were employed in the oldest of map uses in Canada, namely the placing of settlers on homesteads in wilderness country. Today Quebec produces a 1:125,000 road map series that will be described in Chapter 13. The Prairie provinces had no need for two-mile maps; their three-mile series provided their mapping needs. British Columbia, like Ontario, had used two-mile maps here and there to support geological exploration, but it was not until 1912 that they started to publish a two-mile series called the degree series, thus named because each sheet covered one degree in both latitude and longitude. Table 9 lists the sheets produced, and it will be noticed that the First World War seriously interfered with production.

Another two-mile series, with a completely different style, was started in 1908. This was the military two-mile series published by the Survey Division, General Staff. In 1903 Major Hills, the War Office mapping expert (who had been invited to Canada to report on military mapping) recommended the two-mile series as a worthy compromise between the detailed one-inch maps and the much less expensive reconnaissance maps at four miles to

Table 9 Sheets of the British Columbia Degree Sheet Series

Sheet no.	Mid latitude	Mid longitude	Sheet name	Edition 1	Edition 2	Edition 3
4A	49°30′	118°	Rossland	1912	1926	
4B	49°30′	117°	Nelson	1946		
4C	49°30′	116°	Cranbrook	1912	1936	
4D	49°30′	115°	Fernie	1913	1949	1957
4E	50°30′	115°	Upper Elk River	1913	1925	
4F	50°30′	117°	Lardeau (Duncan River)	1913	1947	1956
4G	50°30′	116°	Windermere	1914	1943	1956
4H	50°30′	118°	Arrowhead	1926		
4J	50°30′	119°	Vernon	1921		
4K	49°30′	119°	Kettle Valley	1921	1923	
4L	50°30′	122°	East Lillooet	1926		
4M	50°30′	121°	Nicola Lake	1927		
4N	49°30′	120°	Penticton	1930	1936	
4P	49°30′	122°	Lower Fraser Valley	1946		
4Q	49°30′	121°	Hope-Princeton	1939		

	122°	121°	120°	119°	118°	117°	116°	115°
51°–50°	East Lillooet	Nicola Lake	Not published	Vernon	Arrowhead	Lardeau	Windermere	Upper Elk River
50°–49°	Lower Fraser Valley	Hope-Princeton	Penticton	Kettle Valley	Rossland	Nelson	Cranbrook	Fernie

the inch.[1] But with the plane table methods of the time, there was very little saving in surveying at this intermediate scale, so the Canadian General Staff opted for the one-inch series. These two-mile maps of 1908 were each derived from four published one-inch maps. The British Columbia series, on the other hand, consisted of exploratory maps drawn from original surveys.

In 1925 the first version of the National Topographic System was unveiled, and the two-mile series was one of the map scales incorporated into the system. This gave the two-mile scale a new respectability, and maps at this scale, with sheetlines and format designed to fit the NTS, were produced by both federal and provincial mapping agencies. Generally, before the Second World War it was used as a derived scale by the military and as an exploratory scale by other agencies who used it where the terrain imposed too many difficulties for one-inch mapping.

After the Second World War the two-mile series became something of an orphan in federal mapping agencies. Almost the entire extent of their resources was funneled into two other scales, the four-mile reconnaissance mapping and the one-inch resource development mapping. The few two-mile sheets published after 1946 were all produced by derivation from the one-inch and later the 1:50,000 maps. But interest was kept alive in the provinces. As will be seen in Chapter 13, British Columbia publishes beautifully contoured maps of the mountainous areas of the province, and Ontario introduced a planimetric series at the two-mile scale for parts of northern Ontario.

In 1957 the last of the federal two-mile sheets was published. This was Goudreau, 42 C/SE, in northern Ontario. For three years there was a pause while the authorities considered the future of the series, then in 1960 the first of the 1:125,000 sheets appeared. This was the Ste-Agathe sheet, 31 J/SE. From then on all revisions of the two-mile sheets came out at the slightly larger scale. Thus 92 I/NE Kamloops Lake appears in 1962 as the fourth edition of the sheet though it was the first version at 1:125,000.

In the 1960s the two-mile maps were popular with two quite different sectors of the map-user public, the forest industry and the tourists. Foresters found these maps good for timber cruising and planning forestry operations, and excellent for forest-fire fighting. In particular, the coordination of water bombing with the fire-fighters on the ground could be organized best on two-mile maps. (The 1:50,000 required too many maps for use either on the ground or in the planes, while the 1:250,000 did not have enough terrain detail.)

Tourists liked the maps for much the same reason. They obtained a good spread of country and sufficient map information from each map. The sheets were never published with a UTM grid as neither the foresters nor the tourists wanted it. To make the map more attractive for the tourists a number of the sheets were published with hill shading. For each sheet there was also an unshaded version.

A number of maps in Quebec were put together very quickly in 1967 and 1968 at the request of the Quebec Forest Protection Agency. These maps can be recognized because they are simply 1:50,000 sheets photographically reduced and published without generalization. They served the purpose for which they were drawn, but many complaints were received from general users who found some of the type almost impossible to read. They also had difficulty with some of the symbolization (roads, marsh symbols, etc.) which became illegible at the smaller scale. These maps were given a special 'forest protection' grid which is used in Quebec.

Tables 10 and 11 list the 1:126,720 and 1:125,000 sheets that have been produced by the federal government.

Table 10 List of 1:126,720 Sheets[2]

Stock number	Name	Area	Style	Edition
031-PSE	Latuque	QUE		48·02
032-CNE	Cuvillier	QUE		56·01
032-ENW	Chabbie	QUE-ONT		51·01
032-ESE	Mistawak Lake	QUE		56·01
032-ESW	Patten River	QUE-ONT	Provisional	51·01
041-ISW	Espanola	ONT	Provisional	49·01
041-PNW	Gogama	ONT	Provisional	51·01
042-ASE	Kirkland Lake	ONT	Provisional	49·01
042-ASW	Timmins	ONT		48·01
042-BNE	Elsas	ONT	Provisional	54·01
042-BNW	Fire River	ONT	Provisional	56·01
042-BSE	Folfyet	ONT	Provisional	54·01
042-CNE	Kabinakagami Lake	ONT	Provisional	56·01
042-CNW	White River	ONT	Provisional	56·01
042-CSE	Goudrfau	ONT		57·01
042-CSW	Pukaskwa River	ONT		56·01
042-ENE	Longlac	ONT		53·01
042-ENW	Jellicoe	ONT	Provisional	52·01
042-ESW	Roslyn Lake	ONT		53·01
042-GSE	Kapuskasinc	ONT		55·01
042-HNE	Montreuil Lake	ONT		51·01
042-HSE	Little Abitibi	ONT		51·01
042-ISW	Coral Rapids	ONT	Provisional	46·02
052-ASW	Fort William	ONT	Provisional	47·02
052-FSW	Rowan Lake	ONT	Provisional	66·01
063-LNW	Rauendale	SSK		49·01
063-ESE	Pasquia	SSK		43·01
063-ESW	Arbourfield	MAN		48·01
082-JNW	Palliser-Kananakis	ALT BC		23·01
082-LNW	Shuswap	BC		55·03
082-LSW	Vernon	BC		51·01
082-NNE	Mistaya	BC		50·02
082-NSE	Yoho	ALT BC		52·02
082-NSW	Glacier Park	BC		74·04
083-GNE	Lac Ste Anne	ALT		53·01
083-GNW	Chip Lake	ALT		52·01
083-GSE	Warburg	ALT		52·01
083-GSW	Pembina	ALT		52·01
083-MNE	Rycroft	ALT		48·01
083-MNW	Blueberry Mountain	ALT		48·01
083-MSW	Bracerlodge	ALT		48·01
083-NNE	McLennan	ALT		48·01
083-NNW	Watino	ALT		49·01
084-OSW	Grimshaw	ALT		49·01
084-DSE	Hines Creek	ALT		50·01
084-DSW	Cherry Point	ALT		51·01
092-ISE	Merritt	BC		50·01
093-PNE	Dawson	BC		51·01
093-PNW	Moberly Lake	BC		51·01
094-ANE	Rose Praire	BC		51·01
094-ANW	Blueberry	BC		51·01
094-ASE	Ft St John	BC		53·01
094-ASW	Hudson	BC		53·01

Table 11 List of 1:125,000 Sheets[3]

	Stock number	Name	Area	Style	Edition
	011-KNE	Cape Breton Highland	NS		70·05
	011-LSE	Georgetown	PEI		71·01
SR	011-LSE	Georgetown	PEI	Shaded relief	71·01
	011-LSW	Charlottetown	PEI		71·01
SR	011-LSW	Charlottetown	PEI	Shaded relief	71·01
	021-HNE	Amherst	NS-NB		69·01
SR	021-HNE	Amherst	NS-NB	Shaded relief	69·01
	021-INE	Cascumpec Bay	PEI		71·01
SR	021-INE	Cascumpec Bay	PEI	Shaded relief	71·01
	021-MNE	Clermont	QUE		69·04
SR	021-MNE	Clermont	QUE	Shaded relief	70·01
	021-MSE	Beaupre	QUE		70·01
SR	021-MSE	Beaupre	QUE	Shaded relief	70·01
	022-CSW	Trois Pistoles	QUE		69·01
SR	022-CSW	Trois Pistoles	QUE	Shaded relief	69·01
	022-DNW	Alma	QUE		71·01
SR	022-DNW	Alma	QUE	Shaded relief	71·01
	022-DSE	Port Alfred	QUE		69·01
SR	022-DSE	Port Alfred	QUE	Shaded relief	69·01
	031-DNE	Bobcaygeon	ONT		65·02
SR	031-DNE	Bobcaygeon	ONT	Shaded relief	65·02
	031-DNW	Orillia	ONT		65·04
SR	031-DNW	Orillia	ONT	Shaded relief	65·04
	031-ENE	Algonquin	ONT		65·02
SR	031-ENE	Algonquin	ONT	Shaded relief	65·02
	031-ENW	Sundridge	ONT		64·04
SR	031-ENW	Sundridge	ONT	Shaded relief	64·04
SR	031-ESE	Haliburton	ONT	Shaded relief	65·02
	031-ESW	Muskoka	ONT		65·05
SR	031-ESW	Muskoka	ONT	Shaded relief	65·05
	031-FNE	Fort Coulonge	QUE-ONT		63·04
SR	031-FNE	Fort Coulonge	QUE-ONT	Shaded relief	63·04
	031-FNW	Golden Lake	ONT		63·01
SR	031-FNW	Golden Lake	ONT	Shaded relief	63·01
	031-FSW	Bancroft	ONT		64·01
SR	031-FSW	Bancroft	Ont	Shaded relief	64·01
	031-GNE	Lachute	QUE		65·04
SR	031-GNE	Lachute	QUE	Shaded relief	65·04
	031-GNW	Buckingham	QUE-ONT		65·05
SR	031-GNW	Buckingham	QUE-ONT	Shaded relief	65·05
	031-JNE	L. Ascension	QUE	Provisional	65·03
	031-JNW	Mont Laurier	QUE	Provisional	67·04
SR	031-JNW	Mont Laurier	QUE	Shaded relief	67·04
	031-JSE	Ste-Agathe	QUE		60·01
SR	031-JSE	Ste-Agathe	QUE	Shaded relief	60·01
	031-JSW	Maniwaki	QUE		67·03
SR	031-JSW	Maniwaki	QUE	Shaded relief	67·03
	031-KNE	Tomasine	QUE		67·04
SR	031-KNE	Tomasine	QUE	Shaded relief	67·04
	031-KNW	Lac Dumoine	QUE		70·02
	031-KSE	Gracefield	QUE		67·04
SR	031-KSE	Gracefield	QUE	Shaded relief	67·04
	031-KSW	Deep River	QUE-ONT		65·01
SR	031-KSW	Deep River	QUE	Shaded relief	65·01
	031-LNE	Lac Beauchene	QUE		66·01
	031-LNW	Tomiko	QUE-ONT	Provisional	65·02
	031-LSE	Mattawa	QUE-ONT		67·02
SR	031-LSE	Mattawa	QUE-ONT	Shaded relief	67·02
	031-LSW	North Bay	ONT		66·02

	Stock number	Name	Area	Style	Edition
SR	031-LSW	North Bay	ONT	Shaded relief	66·02
	031-MNE	Lac Simard	QUE	Provisional	70·01
	031-MNW	New Liskeard	QUE-ONT	Provisional	68·01
	031-MSE	Belleterre	QUE	Provisional	66·01
	031-MSW	Haileybury	QUE-ONT	Provisional	65·02
	031-NNE	Vimy	QUE	Provisional	67·03
	031-NNW	Grand L. Victoria N	QUE	Provisional	54·02
	031-NSE	Cabonga Resevoir	QUE	Provisional	65·03
	031-NSW	Grand L. Victoria S	QUE	Provisional	65·03
	031-ONE	Parent	QUE	Provisional	68·02
	031-ONW	Choquette	QUE	Provisional	68·02
	031-OSE	Kempt Lake	QUE	Provisional	68·03
	031-OSW	Petawaga	QUE	Provisional	65·04
	032-BSE	Barrage Gouin	QUE	Provisional	68·01
	032-BSW	Oskelaneo	QUE	Provisional	68·03
	032-CSE	Lac Faillon	QUE	Provisional	68·03
	032-CSW	Senneterre	QUE	Provisional	47·02
	032-DNE	Taschereau	QUE		72·03
	032-DSE	Rouyn Lake	QUE	Provisional	68·03
	032-DSW	Rouyn-Larder Lake	QUE-ONT	Provisional	68·03
	041-HNE	Byng Inlet	ONT		68·04
SR	041-HNE	Byng Inlet	ONT	Shaded relief	68·04
	041-HSE	Parry Sound	ONT		68·04
SR	041-HSE	Parry Sound	ONT	Shaded relief	68·04
	041-PSE	Maple Mountain	ONT		72·02
	052-ENW	Falcon Lake	ONT-MAN		69·01
SR	052-ENW	Falcon Lake	ONT-MAN	Shaded relief	69·01
	052-LNW	Manigotagan Lake	ONT-MAN		70·01
SR	052-LNW	Manigotagan Lake	ONT-MAN	Shaded relief	70·01
	052-LSW	Eaglenest Lake	ONT-MAN		70·01
SR	052-LSW	Eaglenest Lake	ONT-MAN	Shaded relief	70·01
SR	062-ENE	Moose Mountain	SSK	Shaded relief	67·01
	062-FSE	Turtle Mountain	MAN		70·01
SR	062-FSE	Turtle Mountain	MAN	Shaded relief	70·01
	062-INE	Pine Falls	MAN		70·01
SR	062-INE	Pine Falls	MAN	Shaded relief	69·01
	062-JNE	Lundar	MAN		71·01
	062-JNW	McCreary	MAN		71·01
SR	062-JNW	McCreary	MAN	Shaded relief	70·01
	062-JSE	Gladstone	MAN		70·01
	062-JSW	Neepawa	MAN		70·01
	062-KNE	Oakburn	MAN		70·01
SR	062-KNE	Oakburn	MAN	Shaded relief	70·01
	062-KNW	Russel	MAN		72·01
SR	062-KNW	Russel	MAN	Shaded relief	72·01
	062-KSW	Moosomin	MAN-SSK		70·01
	062-MSW	Ituna	SSK		72·01
	062-NNE	Duck Mountain	MAN		68·01
SR	062-NNE	Duck Mountain	MAN	Shaded relief	68·01
	062-NNW	Kamsack	MAN-SSK		69·01
SR	062-NNW	Kamsack	MAN-SSK	Shaded relief	67·01
	062-NSE	Valley River	MAN		70·01
SR	062-NSE	Valley River	MAN	Shaded relief	70·01
	062-NSW	Roblin	MAN-SSK		70·01
SR	062-NSW	Roblin	MAN-SSK	Shaded relief	70·01
	062-ONE	Lake St Martin	MAN		72·01
	062-ONW	Winnipegosis	MAN		72·01
	062-OSE	Ashern	MAN		72·01
	062-OSW	Ste Rose Du Lac	MAN		70·01

	Stock number	Name	Area	Style	Edition
	063-BNE	Ashmall Point	MAN		72·01
	063-CNE	Pelican Rapids	MAN		72·01
	063-CNW	Red Deer Lake	MAN-SSK		72·01
	063-CSE	Cowan	MAN		72·01
	063-CSW	Woody River	MAN-SSK		72·01
	063-DNE	Etomami	SSK		72·02
	063-DNW	Mistatim	SSK		72·01
	063-DSW	Nut Mountain	SSK		72·01
	063-FNW	The Pas	MAN-SSK		72·01
	063-GNE	Limestone	MAN		72·01
SR	063-KNW	Flin Flon	MAN	Shaded relief	72·01
	072-ENW	Forty Mile Coulee	ALT		72·01
	072-ESE	Pakowki Lake	ALT		72·01
	072-ESW	Etzikom Coulee	ALT		70·01
	072-FNW	Maple Creek	SSK		70·01
SR	072-FNW	Maple Creek	SSK	Shaded relief	70·01
	072-JNW	Lucky Lake	SSK		71·01
SR	072-JNW	Lucky Lake	SSK	Shaded relief	71·01
	072-KNW	Leader	SSK		70·01
SR	072-KNW	Leader	SSK	Shaded relief	70·01
	073-BNW	Redberry Lake	SSK		72·01
	073-BSW	Biggar	SSK		72·01
	073-CNW	Manito Lake	SSK		72·01
	073-GSE	Shellbrook	SSK		72·01
	073-GSW	Witchekan	SSK		72·01
	092-INE	Kamloops Lake	BC		62·04

At present (1979) the federal series is dormant. Only the 47 shaded relief maps listed in Table 11 are being revised and no new sheets are being drawn. The other sheets of the series are still being sold but as stocks run out they will not be replaced.

At the peak of production, 164 sheets existed in the two series. In April 1979 150 were still in stock as follows:

1:126,720	Standard	40
	Provisional	13
1:125,000	Standard	76
	Provisional	21
	Shaded relief	47

The shaded relief maps are in each case also published as standard sheets. The symbolization of the standard version is much the same as that on standard 1:50,000 maps except at the smaller scale fewer buildings are shown and farmsteads are represented by a single building symbol. Smaller topographic features such as ponds, small marshes, etc. are omitted.

The provisional sheets are those that were compiled for the forest protection agencies simply by reducing four 1:50,000 to scale and publishing them with a very simple marginal layout.

Figure 15 is a section taken from the shaded relief version of 31 D/NW Orillia. Tables 10 and 11 list the sheets of the two series that are available in 1979 from the Canada Map Office.

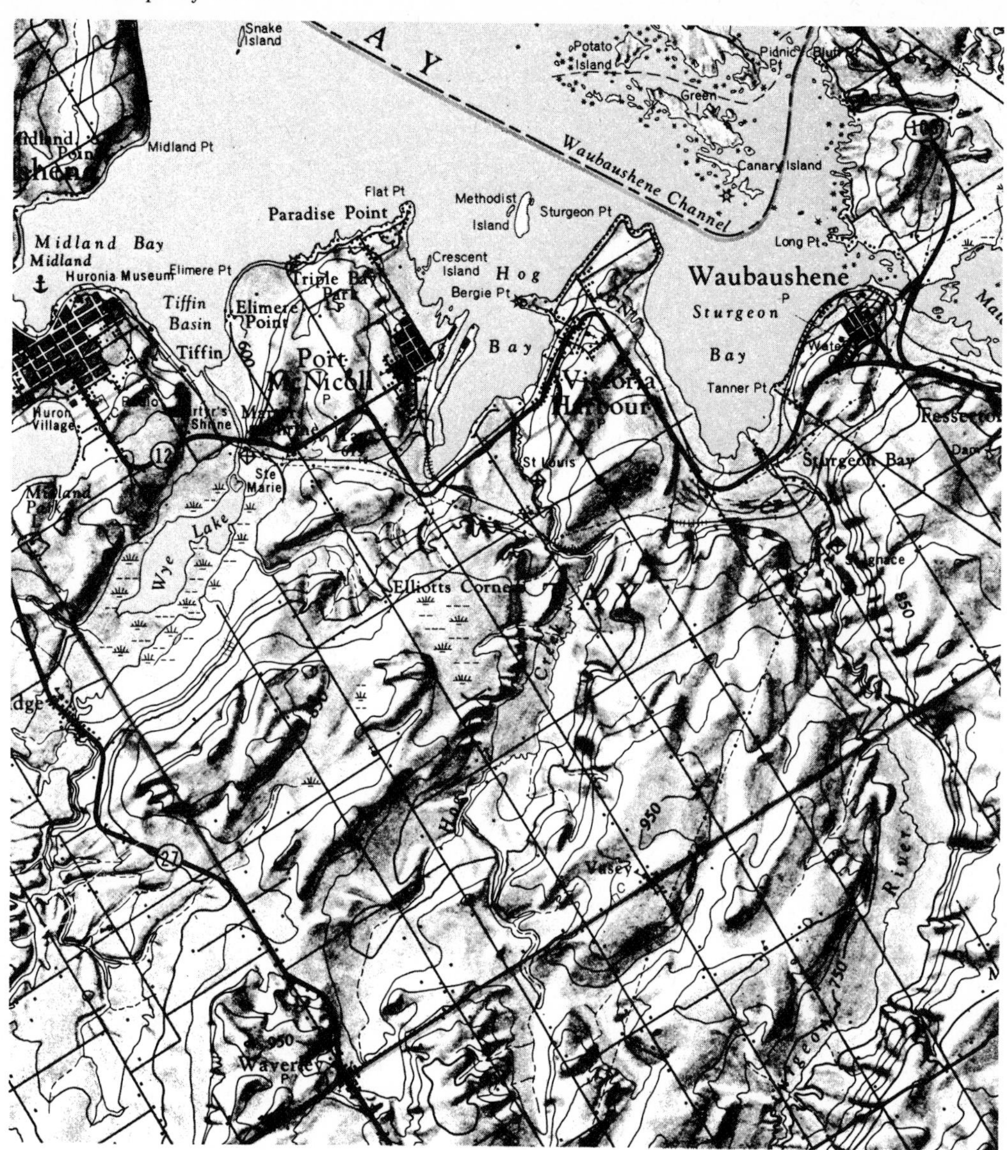

Figure 15 Section from 1:125,000 sheet

7. The Four-Mile Series and the 1:250,000 Series

The Four-Mile Series (1:253,440)

The four-mile series was the third of the basic scales of the original National Topographic System drawn up by the Board of Topographical Surveys and Maps in 1923. The sheet lines were to be fitted into those of the 1:1,000,000 International Map of the World and the first three sheets of the four-mile series which were published in 1926 did conform to this system. However, in 1927 the system of sheet lines was changed to that essentially in use today and subsequent maps in the four-mile series were numbered in the new system.

Seven methods were employed in making the four-mile maps. They are summarized in Table 12, and are described in detail below.

Table 12 Methods of Making the Four-Mile Maps

Method	Terrain in which used	Period in use
1 Derivation from previously produced maps at larger scale.	All types	At all times
2 Ground photo-topographic methods	Mountainous areas	Pre-1939
3 Aerial—oblique plotting.	Forested flat areas	Pre-1939
4 Trimetrogon plotting.	All types	1945–1947
5 Vertical photography with triangulation for horizontal control and photo-topographic methods for vertical control.	Mountainous areas	1947–1954*
6 Vertical photography with Shoran horizontal control and APR (air profile recorder) vertical control.	All types of Arctic terrain	1952–1957*
7 Vertical photography with tellurometer horizontal control and either barometer or APR vertical control.	All types	1957–1968*

* Fieldwork done after 1950 was used for 1:250,000 mapping.

Derivation From Previously Produced Map Series

Some sheets were derived from existing one-inch maps for parts of southern Ontario and some from the three-mile map series. It has already been pointed out that one method used in Canada to produce medium-scale topographic mapping was to use the surveys of land subdividers for horizontal control, and to add contours and other topographic detail by subsidiary surveys as was done in the case of the Prairie three-mile maps. In the early 1930s some of this mapping was converted to the four-mile scale. The contour plate was reduced photographically without alteration. The drainage plate was also reduced photographically, but the muskeg symbol (a small blue coniferous tree) was replaced by the normal swamp symbol after 1939. The green plate, which on the three-mile maps showed a pattern of tree symbols, was changed to a solid light-green area tint. There was very little change in the depiction of the road network in the conversion to the smaller scale. The style of type was altered to conform with the four-mile specifications, but the number of place names in a given area was almost identical in the two scales. The depiction of man-made features, too, survived the scale reduction without any apparent generalization. The individual farmhouses, which were rigorously shown on the three-mile maps, also appear on the four-mile. It is quite surprising to find such detail at four miles to the inch, but the overall effect is quite pleasing, giving a very good indication of the occupation of the country.

Ground Photo-topographic Techniques

In the mountainous regions of the west a fourth-order triangulation net was run through the area with sides of triangles ranging from six to twelve miles in length. A panorama of photographs was taken from each station and from these additional control points were intersected and elevations were calculated from vertical measurements on the photographs. The topographer could then sketch in additional detail from the pictures much as he would on a plane-table in the field.[1]

During the development of the photo-topographic work it became the practice to rule a perspective grid onto a print of the photograph to assist the topographer in positioning horizontal detail (Figure 16). This device became most important in the next stage of Canadian four-mile mapping, which was the development of a mapping system that would cope with the relatively flat, forested areas of Canada's north. A wide field of view was essential to the system, and this required an elevated camera station. The photographic aircraft was the obvious solution.

It must be mentioned that the Topographic Division of the Geological Survey also used the ground photo-topographic technique for mapping in areas where the depiction of the geological structures was best served by maps at this scale. It has already been pointed out that geologists have their own priorities in depicting topography. They do not show forest cover, they employ a very simple set of road symbols, and they tend to ignore the works of man (isolated houses, barns, etc.) unless they are connected in some way with mineral development. The four-mile sheets prepared by the Geological Survey exhibit this geological bias, and hence are quite different from those of the Military or Topographical Surveys, and in exactly the same way that their one-inch sheets differed. But these sheets did depict the topography, and there was no thought of the Topographical Survey redrawing them just for the sake of uniformity. A few of them even survived the scale change to

1:250,000 in the 1950s. The Geological Survey did not use any of the other methods of four-mile mapping which will now be described.

Aerial Oblique Plotting[2]

In 1925, the Vickers Vedette came into production and several were purchased by the Royal Canadian Air Force. These craft were ideally suited both to take oblique aerial photographs and to operate from the small lakes that dot the north. The perspective grid that had been developed for mountain work was directly applicable to the plotting from aerial obliques. Photographs were taken, looking forward and 45° to both sides from the open cockpit of the Vedette while flying at about 5000 feet. A record of the altimeter was kept at the time of each exposure, and by knowing the approximate vertical tilt of the camera and the focal length, a perspective grid could be drawn. Scale and position could be computed if two ground control points appeared in the picture, and this scale could be carried forward some distance from one photograph to another.

Flights were planned systematically in parallel straight lines at right angles to stadia traverses run along water courses at about 100-mile intervals. In very remote or inaccessible country even these traverses were dispensed with, and whole four-mile sheets were compiled with only a skeleton framework of astro-fixes as control.

The maps produced by aerial oblique plotting proved invaluable during the detailed exploration of the Canadian Shield and the wooded areas of the interior plains where the range of elevation is small. For example the extreme variations in elevations over the

Figure 16 Ground photograph taken with a photo-topographic camera showing perspective grid for drawing topographic detail

entire north–south portion of the Manitoba–Ontario boundary, which crosses some very rugged country, is only 200 feet. The thousands of small lakes of this region greatly assisted the work. They were easily identifiable in the plotting, and they provided 'airfields' for the flying-boats during the field work phase. In fact, the first three four-mile sheets to be published were for this area—Lac Seul (present 52K), Pointe du Bois (present 52L) and Carroll Lake (present 52M).

In 1934 work on the four-mile series was seriously curtailed through lack of funds. The ground photo-topography continued in the mountains of British Columbia, but elsewhere the work came to a stand-still. In 1939 the country went onto a war economy and the four-mile project was shelved so that all personnel could work on the production of eight-mile aeronautical charts which were needed for defence requirements.

Trimetrogon Photography

After World War II the Canadian mapping authorities were faced with a completely new set of objectives. Wartime activities in the Canadian Arctic, especially the European and Russian airlift operations, had focused the attention of many Canadians on the far north. Wartime training had provided a large cadre of mapping specialists in all trades, and photographic aircraft were available, with crews to man them. During the war the US Air Force had developed the Trimetrogon mapping system employing three cameras which photographed, simultaneously, the complete panorama from horizon to horizon at right angles to the line of flight of the aircraft. This method, which will be described in more detail in Chapter 8, was instituted on a 'crash programme' basis in many countries when the USAF needed the largest possible photographic cover, in the shortest possible time, for the production of the 1:1,000,000 World Aeronautical Chart series. During the last year of the war the RCAF assisted in this programme and continued it into 1946 and 1947. By 1948 the entire land area of Canada which had not previously been photographed with vertical photography was covered with 20,000-foot Trimetrogon photography flown along lines 16 miles apart. This was mainly used to improve the Canadian eight-mile aeronautical charts, but in 1947 six planimetric four-mile sheets were drawn. However, planimetric maps were on the way out in Canada and as the Trimetrogon method could not produce reliable height information it was dropped.[3]

Vertical Photography with Triangulation and Ground Photos

In mountainous areas the general availability of vertical photography provided an improvement on the pre-war photo-topographic method. Ground photographs were still employed but only to provide height control and a few additional intersected horizontal control points. The contouring was done by sketching on prints of the vertical photography while they were being examined under a desk stereoscope. Radial intersection frameworks[4] were used to provide horizontal control between the field survey points. This system was used until 1955.

Vertical Photography and Shoran Control

In 1947 Canada's first serious long-range mapping programme was approved. With its built-in contingency for an annual review to accommodate changing conditions, it had a most profound effect on the future mapping of Canada. The first such far-reaching effect resulted from the initial concept that it was necessary to extend both geodetic control and vertical photography over all of Canada at the earliest possible moment so that in an emergency the 1:50,000 map of any area could be produced without the normal delay imposed by seasonal conditions.

The second basic decision resulting from this long-range plan was that the four-mile mapping of Canada should be completed within a 20-year period. This scale was chosen because at the time the imperial scales were still in effect and the four-mile-to-the-inch scale was the largest that could be completed within the agreed limit of 20 years.

Geodetic horizontal control was the first requirement, but any plan to extend by conventional geodesy over the huge area of Canada was out of the question. The various techniques that had been developed during the war for bomber-guidance were examined, and of these the Shoran control system was considered most efficient for the precise measurement of distance. In 1947 this system was incorporated into a trilateration programme covering the whole of northern Canada. By 1957 this had resulted in a complete net of triangles with sides about 300 kilometres long and, with an estimated length error of less than six metres, the relative accuracy was about 1:50,000.

At the same time as the Shoran net was being developed, vertical airphoto coverage was being extended. The RCAF assisted in this work from 1947 to 1949, but from 1950 onward virtually all of it was done by private companies. By 1957 vertical photo coverage of Canada was complete.

Originally a system employing Shoran-controlled photography was used in which lines of photography were flown in an east–west direction spaced at 20 minutes of latitude and in a north–south direction at 80 minutes of longitude. A reading was taken of the position of the aircraft at the moment of exposure of each photograph, and the positions of the ground plumb-points were found by levelling the photogrammetric models using any available vertical control. The result was a grid of known horizontal positions along lines spaced about 40 miles apart in an east–west direction and about 23 miles apart north to south.

Vertical Photography and Tellurometer Surveys

The Shoran photography method of horizontal control had many defects. It was expensive, and its accuracy was open to question because of the difficulty in obtaining position values on the ground from Shoran fixes on the aircraft some 20,000 feet in the air. In addition no identifiable marks were left on the ground for subsequent survey work. It was discontinued when the tellurometer became available in 1957. All the traversing done on islands of the Arctic was carried out by parties equipped with tellurometers. It should be mentioned that bad closures on the Shoran ground stations were surprisingly few, a fact which reinforced confidence in the basic net.

Another important advantage, for the quick despatch of mapping in the Canadian Arctic islands, is the sea, which provides a ready and ubiquitous vertical datum. On the islands vertical control was established by air profile recorder lines run from sea level to sea level. On the mainland, far from the sea, the situation was slightly more difficult. In certain areas helicopter-borne barometer traverses were run between third-order level lines which had been set out over the frozen tundra during the winters or along rivers during the

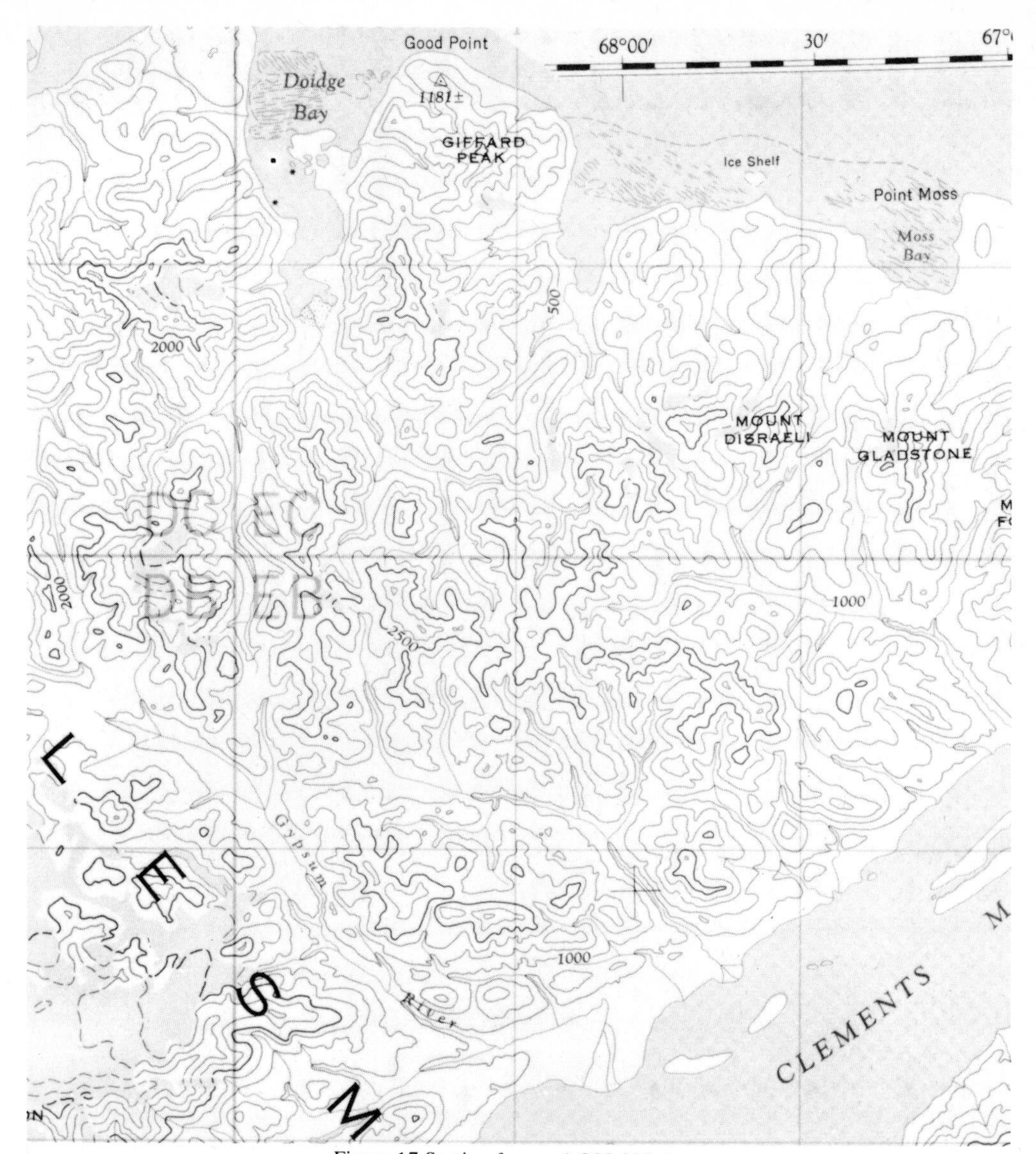

Figure 17 Section from a 1:250,000 sheet

summers. As can be imagined, the winter surveying was extremely difficult work carried out in a brutal climate, but most often high standards of accuracy were maintained. An example of this type of work is the line run from Great Bear Lake to tidewater in Franklin Bay which had a closure of just under three feet. In other areas the air profile recorder was used with a system of cross-flights being employed to ensure the required accuracy. Contouring was done either with the desk-stereoscope or by multiplex.

The 1:250,000 Series

In 1948 the military mapping agencies of the United Kingdom, the USA and Canada agreed upon the desirability of changing from duodecimal scales to round-figured representative fractions and in 1950 the civilian mapping organization in Ottawa concurred.[5] The four-mile series was therefore gradually converted to the 1:250,000 series. In southern areas many sheets could be derived from information drawn at larger scales. Revision work was also pressed, as many of the four-mile sheets became obsolete due to the rapid growth of urban areas and transportation nets.

The symbolization used on the 1:250,000 maps has a superficial resemblance to that of the 1:50,000 series but because of the five-times reduction in scale there are important differences. The original specifications of the 1:250,000 series have been revised on several occasions to improve the presentation or to cope with new conditions. When the Arctic sheets of the series were first compiled a new set of symbols was required to depict such features as pingos, tundra polygons, ice shelves and string bogs. In the temperate latitudes changes were at times made to facilitate revision in an effort to keep the maps of the series up-to-date so that some symbols have been dropped completely while others disappeared and have now returned. This is clear from a detailed examination of the more important symbols.

Principal Symbols[6]

BUILDINGS. In the early days of the 1:253,440 series it was inevitable that these maps would be compared with the sheets of the three-mile series. It would have been disastrous if the newer series showed substantially less topographic information than the older. As the three-mile maps faithfully depicted farmsteads, isolated barns and other rural structures, the early sheets did so as well. Following World War II the competition from the three-mile series faded and the 1:50,000 series took over the role of depicting rural settlement. In 1950, when the scale was changed from 1:253,440 to 1:250,000, it was decided to depict only landmark buildings (this included churches and schools) and railway stations. This policy was continued until 1965 when churches and schools in settled areas were deleted. The symbol E, which for many years indicated the position of that most useful Prairie landmark, the grain elevator, has also been dropped.

On wilderness sheets, churches, schools, trading posts and RCMP posts are shown, as are large commercial enterprises such as sawmills, fish canneries, etc. The latter are represented by one or two building symbols and an appropriate label.

POPULATED PLACES. Villages with a population up to 500 are shown by an open circle 1.25 mm in diameter. Towns with a population of 500 to 2000 are depicted by an open square with sides 1.8 mm long. Places with a population in excess of 2000 are outlined and shaded with the pink urban area tint. Arterial roads and railways are shown within the tinted area.

OTHER MAN-MADE FEATURES.

Airfields are shown by an aeroplane symbol.

Bridges are shown only if they are large and important.

Camp sites and picnic grounds have been shown since 1972 in order to improve the value of the series to tourists. Camp sites are symbolized by a wigwam silhouette; picnic grounds by the end-view of a picnic table.

Dams and dykes are shown only if they are of substantial size.

Fire lookouts are shown by a 1.25 mm circle with a centre dot and a label.

Oil and gas fields are shown by an outline and a label. Individual wells, which are shown at larger scales, are not depicted.

Pipelines are only shown if they are trunk lines.

Portages are shown using the trail symbol. On early sheets of the series the length of the portage was given in chains, but in the years immediately following World War II portages were not depicted with the previous care because, it was reasoned, the bush plane had replaced the freight canoe. In recent years recreational canoeists have demanded a return of this symbol.

Powerlines are depicted only if they are of high tension and carried on substantial towers.

Radio stations are shown in wilderness areas by a 1.25 mm circle with a centre dot and the label 'Radio'.

Railways are depicted in the same manner as on the 1:50,000 series except that railway yards, repair shops, etc. are not shown.

Ski lifts are shown but only if over 1.5 km long.

Survey markers, both horizontal control and bench marks, are shown in remote areas if they are monumented. The former is depicted by an equilateral triangle with sides 1.5 mm, the latter by a broad arrow symbol and a number indicating the elevation in feet or metres depending on the unit used for the contours on the sheet.

Telephone lines in rural areas are now so commonplace that there is no point in showing them. Telegraph offices have also lost their importance in this day of long-distance phone calls.

Telephone trunk lines are also rapidly disappearing because of the increased efficiency of microwave transmission, but where they do exist in wilderness areas they are shown by a row of small inverted T's.

Towers of all kinds (microwave, observation, radio, television, etc.) are shown by the circle and dot symbol with an identifying label and the height of the tower if it is over 30 metres.

ROADS. The road depiction is essentially the same as that on the 1:50,000 except that due to the reduced scale many of the lower classes of roads are omitted if their depiction would cause a congestion of map detail. In the early days orange was used to indicate secondary roads but this practice was discontinued after World War II. However it was reinstated in 1967 when it was adopted for gravel and dirt roads. Cased roads were dropped in the same year.

HYDROGRAPHY. Lakes, rivers and streams are depicted with the same symbolization as they are on the 1:50,000 maps. The following minimum dimensions* are observed:

Streams	10,000′ in length
Double line streams	350′ in width
Lakes and ponds	1000′ in diameter

VEGETATION. Wooded areas are shown by a green tint. Woods smaller than 2000 feet in diameter are not shown and neither are clearings in forests that are smaller than that figure. Marshes (open wet ground) and swamps (wooded wet ground) are shown by the usual bulrush symbol if they are over 2000 feet in diameter. On early sheets of the 1:253,440

* See Appendix 2 for metric minimum dimensions.

series a blue coniferous tree symbol was used to depict muskeg, but unfortunately this typically Canadian symbol was dropped in 1939.

CONTOURS. The contour symbols are the same as those used on the 1:50,000 series with the following contour intervals:

Terrain type	*Imperial (feet)*	*Metric (metres)*
Very flat	100	20
Normal	100	50
Hilly	200	100
Mountainous	500	200

On some of the early 1:253,440 sheets, when the series was competing with the three-mile series, contour intervals of 25 and 50 feet were used in flat areas of the Prairies. When the series was converted to 1:250,000 the smallest vertical interval was set at 100 feet. As seen above, the metric sheets of flat areas will return somewhat to the better ground depiction of pre-World War II days.

CADASTRAL AND BOUNDARY INFORMATION. This series is not intended to show cadastral information below township boundaries, but on the Prairies the DLS pattern is so regular that a fine grey line is used to depict surveyed section lines when they do not fall along roadways. International, provincial, county and township boundaries are shown using the same symbols as those on the 1:50,000 series.

MARGINAL INFORMATION. The data in the margins of the 1:250,000 series is almost identical to that on the 1:50,000 series. The information in the margin has been bilingual since 1960, but at the time of writing the map labels are still in English only. In 1967, with adoption of the orange road symbol, a minor modification was made in the surround layout. The legend on the front of the map was reduced to simply the road symbols, which are demanded by the military. An extensive legend was printed in grey on the back of the map. A rectangular four-mile square point-location grid was carried on the original planimetric sheets, but it was discontinued after World War II. It reappeared in 1965 in the form of the light blue 10,000-metre squares of the UTM grid.

The completion of the series received added impetus in 1961. The RCAF developed a new low-level navigation technique that required a new style of air chart at a scale of 1:250,000. They proposed to practise this type of flying northeast of Edmonton and required contoured 1:250,000 maps of a large area as quickly as possible. Fortunately, much of the area had already been mapped but a rather extensive region comprising 11 sheets surrounding Lac La Ronge had not been touched. As the federal agencies were already completely committed, this work was turned over to a private company. The specifications were written in May 1961, and the manuscripts were completed in June 1962. Another portion of unexpected assistance came from the Canadian petroleum industry. In 1963 several oil companies became interested in the oil potential of the Arctic islands. They urgently needed one-inch maps of several of the western islands and proposed that a joint government-industry project be launched to do this mapping. The resulting sheets were suitable both for the oil geologists and as input for the 1:250,000 programme.

By 1967 (the end of the 20-year 'target period') all sheets were completed except those covering three islands in Hudson Bay: Coats Island (45J), Mansel Island (35L) and Nottingham Island (35 M-N). These all appeared in 1970.

Revision

The work on the 1:250,000 series will of course never be complete as there will always be a need for revision. The revision of this series of 918 maps covering half a continent is an operation fraught with difficulties. To make the work of revision manageable Canada has been divided into revision cycle areas. The 10-year cycle area includes the maritime provinces, southern Ontario and Quebec, the Prairies, and the Vancouver, Victoria and Fraser Valley parts of British Columbia, and is covered by 119 maps. The 15-year cycle area covers the remaining portion of the country south of the Wilderness Line shown on Figure 14A. One hundred and eleven maps cover this area. The 30-year cycle area lies north of the Wilderness Line and is covered by 688 maps.

Some additional resources are provided to expedite the revision of certain sheets before they are due for revision if and only if some remarkable change has taken place in the topography, such as the extensive damming of a river or the establishment of a new town. The normal method of revision of a 1:250,000 sheet is to derive the new data from recently revised 1:50,000 sheets, and for this reason every effort is made to revise the larger scale in blocks of the 16 sheets of a 1:250,000 map. In wilderness areas this system is impractical (or impossible if the 1:50,000 coverage has not yet been drawn) and the 1:250,000 sheets are revised directly from new aerial photography. Because of the expense in flying new photo cover, every effort is made to keep track of northern activities. It may come as a surprise to residents of smaller countries that in an open democracy a major government office might not hear of a mine being opened, a major road being opened or even an air-strip being established. But this does happen, and a small 'intelligence group' is kept hard at work searching out such topographic developments. For wilderness sheets over-print revision is employed extensively as such point and linear features (townsites, air-ports, roads, railways, pipelines, etc.) can be added to the map simply by reprinting the map with one additional colour (purple) adding the new features.

The 'milestone' of finishing all first editions of the series was one of considerable importance in Canadian mapping. A complete topographic base map is now available for the innumerable studies that need such a framework. Aerial photography can be indexed on a uniform scale. The toponymy of the whole country can now be regularized on a firm geographical base. Detailed geographical studies can be carried out without running into gaps in the map coverage. The sources of all rivers can be traced; the heights of all mountains are either shown or can be closely estimated from the contours. The list of the uses of the complete series is almost endless.

8. The Eight-Mile Series and Canadian Aeronautical Charts

There is little doubt that the single most important factor in the exploration and development of the Canadian northland was the invention of the aeroplane. Before the arrival of the bush pilot, travel throughout this land was painfully slow and restricted to a few established trails and waterways. With the arrival of the float or ski-equipped aeroplane immediate access to thousands of square miles became available. Distances which before meant a season's strenuous travel became a mere afternoon flight.

At the close of the First World War, Canada had the necessary pilots and aircraft, but, from a mapping point of view, the country was far from ready for this 'invasion' of the Arctic. In 1920 (the year of the first commercial flight into the North) the Canadian mapping agencies were still struggling to map the farmland of southern Ontario, Quebec and the Prairies. It is doubtful if they even gave a second thought to the possibility of deflecting some of this mapping effort toward the production of northern air charts.

It must not be presumed from the foregoing that the Canadian government in 1920 was not interested in air charts; it was simply that the immediate need seemed remote and the resources for their production not available. But there were indications of a desire to get started in this new field of cartography. Canada had become a party to the International Air Navigation Convention in 1919, and in 1920 the Chief Geographer was requested to study the preparation of aeronautical maps.[1] In 1923 Dr Deville, the Surveyor-General, made a detailed study of the prototype 1:200,000 and 1:1,000,000 aeronautical maps that had been produced in Britain and France by working groups of the International Commission for Air Navigation. However, it was not until 1926 that the first experimental air chart was produced in Canada. This was the Winnipeg Aeronautical Map drawn at eight miles to the inch (1:506,880) covering the area around the Winnipeg Air Base.

The development of the Canadian Aeronautical Chart series and the associated small-scale topographic maps will be covered under the following headings:

The eight-mile Topographic and Air Chart series.
The 1:500,000 Topographic and Air Chart series.

The 1:1,000,000 world aeronautical charts.
The 1:250,000 visual terminal area charts.
The military air charts.
(a) Joint operation graphics.
(b) Joint arctic weather station charts.
(c) Low level pilotage charts.
(d) Low level aeronautical charts.

The Eight-Mile Topographic and Air Chart Series

Canada's aeronautical chart policy has from the earliest days been closely connected with the eight-mile (1:506,880) map series. This scale of map was planned by the designers of the National Topographic System to be the second smallest in the range of scales which were originally set at 1, 2, 4, 8 and 16 miles to the inch. No sheets of the 16-mile series were ever drawn, so except for three experimental maps of the International Map of the World 1:1,000,000 series the eight-mile maps were the smallest scale topographic maps published in Canada until after the Second World War. The eight-mile series was a pioneer series with most maps being compiled directly from field surveys and air photographs. The first of the series, 94 SE Hudson Hope, was published in 1929.

It is difficult to determine why the Hudson Hope sheet was chosen to launch the series, but one thing is certain, air navigation had nothing to do with it. Probably the Surveyor-General wanted to bring out the four-mile sheet 94 A which contains the northern portion of the Peace River Block (which had been partially surveyed under the DLS system by British Columbia surveyors, see Figure 12 in Chapter 5), but the survey data for the rest of the area was so sparse that he could not in good conscience publish the data at four miles to the inch. The solution was to publish at the eight-mile scale and label the product 'exploratory'.

In general appearance 94 SE looks surprisingly like the eight-mile aeronautical charts that would follow. The area within the neat-line was given a buff tint that is very close to the shade of the 2000–3000 feet altitude tint that would subsequently be used on the charts. Contours are in brown at 500-foot vertical intervals, but a note warns that they are approximate. The rest of the legend consists of the road classification (automobile road, wagon trail and pack trail), the cadastral information (simply the Interprovincial Boundary and DLS township lines), a village and post office symbol, and the telegraph line symbol. Telephones had evidently not reached the Peace River country in 1929.

One curious feature about this original eight-mile sheet is that it is partly gridded. A system of squares four miles to the side is printed in fine grey lines which are parallel and at right angles to the 124th meridian, which is both the left margin of the sheet and the central meridian of the zone of the 8° Transverse Mercator projection in which 94 SE falls. (See Chapter 15 for more information on this projection.) However, the grid stops abruptly at the boundary of the Peace River Block where the DLS takes over as a reference system. We have on this sheet a preview of the heated argument that would reach a climax some 30 years later when gridded maps were forced on Prairie map-users despite their objections that they did not need or want gridded maps.

In 1929 Canada had started mapping at eight miles to the inch. The following year the Surveyor-General's office issued the Winnipeg to Regina air navigation strip map, also at eight miles to the inch, and Canada was at last into the aeronautical chart business in earnest. In short order more 'air navigation maps' (as they were called until World War II) were produced. This new commitment came at exactly the right time, because in 1928 the responsibility for land surveying in the Prairie provinces had been turned over to the provincial authorities, and although the three-mile maps continued to be produced by the federal Surveyor-General, his surveyors no longer carried out the field work. Thus this new work of preparing small-scale maps for air navigation was received with enthusiasm, particularly as it meant the continued employment of a number of surveyors who otherwise would have had to be released during the economic depression in the 1930s.

Canada's first air chart to be put on sale to the public deserves a description. As has been mentioned, the scale of the Winnipeg to Regina map was eight miles to the inch. Figure 18 shows the legend, but as it is here in black and white, some additional remarks are necessary. Trunk roads are in orange, township boundaries in grey, and railways in black. The drainage, including the marsh symbol, is shown in dark blue while the open water is in a medium blue. (This darker tint than that used on topographic maps makes the small lakes and ponds of the Prairies stand out quite sharply.) The forest is portrayed by tree symbols in green. Altitude tints increase from buff to dark brown in intervals of 500 feet. Buttes and sharp mounds are accentuated by hachures.

The air information, including the power line symbol and the suggested air route, is in red. In 1930 there were of course no radio aids to navigation, so the airport rotating or flashing beacons were the only man-made aids to visual navigation. The strip map is printed in two sections on the sheet, each section 80 miles (10 inches) wide by 192 miles

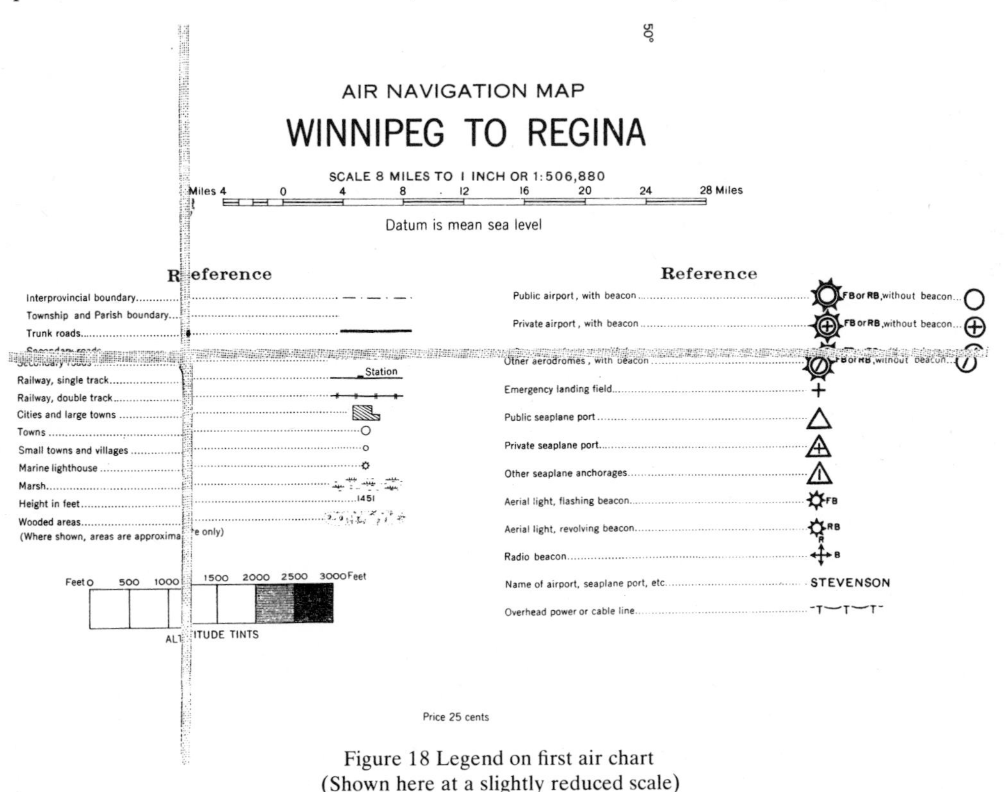

Figure 18 Legend on first air chart
(Shown here at a slightly reduced scale)

(24 inches) long. The back of the sheet carries a profile of the air route and enlarged plans of the airdromes shown on the map.

In 1931 the Department of National Defence published a series of strip aeronautical maps running from Windsor, Ontario to Rimouski, Quebec. The specifications for these were much the same as the Winnipeg–Regina map except that the scale was four miles to the inch (1:253,440). The sheet dimensions were 32 miles (8 inches) wide by 240 miles (30 inches) long. Eight maps were required to cover the route.

Also in 1931 the second eight-mile sheet was published, Winnipeg Lake (63 SE). Thus we have the fortunate occurrence of a map series that was admirably suited as a base for an aeronautical chart series appearing at precisely the moment when the experimental charts had been tested and approved by members of Canada's flying community.

This was still in the darkest months of the economic depression and all expenditures had to be carefully evaluated. The absolute lack of any sort of detailed mapping in the north weighed heavily in favour of the new series, but to get the maximum benefit from the cost of the eight-mile series it had to be designed to perform the dual function of a basic topographic map and an aeronautical chart. The air information would be added as an overprint to the basic map, with both versions being available to the public.

The cartographic detail on the Winnipeg–Regina map had been derived from the three-mile maps, and that of the Windsor–Rimouski set from the one-inch coverage. But the urgent need for navigation maps was in the north where pilots were reduced to using ancient general maps at scales of 50 and even 100 miles to the inch. In 1931 it was decided jointly by the Department of National Defence and the Department of the Interior to make strip maps of the more heavily travelled Arctic air routes. The concept was quite simple. The oblique photography method would be used. The RCAF would provide the aircraft and air crew including the photographer. Fortunately two of the surveyors who worked in the first year of this project, John Carroll and Eric Fry, left detailed reports on how the work was done.[2] A brief resumé of these reports follows.

Each party was equipped with two planes, an open cockpit Vickers Vedette flying boat for the photography and a Bellanca cabin plane for supplies. The party was then assigned a certain number of air routes to survey. The party's task for the season was to produce photos and survey data for about 40,000 square miles of mapping (i.e. the area of two eight-mile sheets). Supplies were obtained locally whenever possible, but often long supply trips had to be made by the cabin plane.

The technique of mapping from oblique photographs required the plane to fly each photographic course in a series of straight-line segments. A section of the area to be photographed would first be reconnoitered by flying over it (without the camera operating) using the best available map. If this map showed too little detail, as was generally the case, the navigator would make a rapid sketch map to help in the subsequent photo flight. Then the route would be reflown in segments using the sketch map to keep on-line. Every 50 to 75 miles a landing would be made on a lake or wide river so that an astronomic fix could be taken. This required a stay of at least one night (often several) so that the surveyor could take readings on a sufficient number of stars to establish the latitude and longitude of the camp. The observing point would be at a place that could be identified on the photographs (often at the shore of a lake on a point of land or beside a stream flowing into the lake) and thus give the portion of the map that would result from the work an accurate scale and orientation. At the end of the season the photographs, the results of the astrofixes and the sketches of the control-point locations were turned over to the Geographical

Section of the General Staff where navigation strip maps were prepared. Table 13 lists the air navigation maps that were produced in this way.

Table 13 Arctic and Sub-Arctic Strip Maps (published between 1931 and 1939)

Name	Location (parts of sheets)	Scale
Churchill River–Fidler Lake	54 E and 54 L	4 miles to 1 inch
Stoney Rapids–Wholdaia Lake	74 P and 75 A	4 miles to 1 inch
Wholdaia Lake–Barrow Lake	75 A and 65 E	4 miles to 1 inch
Barrow Lake–Dubawnt Lake	65 L and 65 N	4 miles to 1 inch
Dubawnt Lake–Beverly Lake	65 N and 66 C	4 miles to 1 inch
Beverly Lake–Lower Thelon River	66 C, 66 B, 66 A	4 miles to 1 inch
Sifton Lake–Thelon River	66 D and 75 P	4 miles to 1 inch
Fort Reliance–Sifton Lake	75 J and K	4 miles to 1 inch
McLeod Bay–MacKay Lake	75 L, 75 M, 76 D	4 miles to 1 inch
Lake Aylmer–Lake Beechey	76 C and 76 G	4 miles to 1 inch
Lake Beechey–Bathurst Inlet	76 G, 76 J, 76 O	4 miles to 1 inch
Lac de Gras–Bathurst Inlet, West Sheet	76 D and 76 E	4 miles to 1 inch
Lac de Gras–Bathurst Inlet, East Sheet	76 L and 76 K	4 miles to 1 inch
Yellowknife River–Reindeer Lake	85 J, 85 O, 86 A	4 miles to 1 inch
Reindeer Lake–Point Lake	86 A, 86 H	4 miles to 1 inch
Point Lake–Big Bend	86 H—86 J	4 miles to 1 inch
Hunter Bay–Coppermine	86 J—86 K	4 miles to 1 inch
Hunter Bay–Dease Bay	86 F—86 L—96 I	4 miles to 1 inch
Dease Bay–Coppermine	86 L—86 N—86 O	4 miles to 1 inch
Rae–Hardisty Lake	85 K O 85 N—86 C	4 miles to 1 inch
Hardisty Lake–Hunter Bay	86 C—86 E	4 miles to 1 inch
Waterways–Fitzgerald	74 D to 74 M	8 miles to 1 inch
Fitzgerald–Providence	74 M to 85 F	8 miles to 1 inch
Providence–Camsell Bend	85 F to 95 J	8 miles to 1 inch
Camsell Bend–Norman	95 J to 96 C	8 miles to 1 inch
Norman–Thunder River	96 C to 106 O	8 miles to 1 inch
Thunder River–Mackenzie Delta	106 O to 107 C	8 miles to 1 inch

During the remaining years of the 1930s, very good progress was made in extending the eight-mile series. As with any dual-purpose creation, some conflicts develop between the requirements of the two classes of users. This series was no exception. The representatives of the pilots pointed out that the names of streams and the location of post offices could certainly be left off. The land-based users found that the 500-foot contour interval gave a very generalized terrain depiction, and they objected to the omission of the smaller streams. A compromise was agreed, and although no one was completely happy the sheets were well used, and new sheets were eagerly awaited. Table 14 lists the sheets of the series produced between 1929 and 1939. It will be noticed that many of these sheets are named after the most important airway on the sheet, for example Cranbrook-Lethbridge.

The Second World War gave great impetus to Canadian aeronautical charting. The Commonwealth Air Training Plan, which was set up in Canada at the outset of the war, required a vast number of charts of southern Canada, while the Northwest Staging Route to Russia and the Crimson Route to Europe extended the chart requirement into the northernmost Arctic. At first the Canadian oblique photography method was pressed into service, but the much superior American Trimetrogon system (so named because it employed three cameras with patented Metrogon lenses) was used. With great assistance from the American Air Corps, vast areas of the Canadian north were photographed.

Table 14 Eight Miles to One Inch Published or in Preparation 1939

1 Charlottetown–Sydney	11 SW (N½) and 11 NW (S½)
2 Fredericton–Moncton	21 SE (N½) and 21 NE (S½)
3 Megantic	21 SW (N½) (a half-sheet)
4 Ottawa–Montreal	31 SE
5 Toronto–Ottawa	31 SW
6 Upper Ottawa River	31 NW
7 Hearst–Cochrane	42 SE in preparation
8 Nakina-Pagwa	42 SW (N½) and 42 NW (S½) in preparation
9 Sioux Lookout–Nipigon	52 SE (N½) and 52 NE (S½)
10 Kenora–Hudson	52 SW (N½) and 52 NW (S½) in preparation
11 English River	52 NW provis. Topo only
12 Brandon-Winnipeg	62 SE (N½) and 62 NE (S½) in preparation
13 Indian Head-Rivers	62 SW air and topo (in preparation)
14 Lake Winnipeg	63 SE provis. Topo only
15 Medicine Hat-Maple Creek	72 SW (N½) and 72 NW (S½)
16 Artillery Lake	Parts of 75 NE and 75 NW
17 Cranbrook–Lethbridge	82 SE air and topo
18 Okanagan–Kootenay	82 SW air and topo
19 Rae	Parts of 85 NE and 85 NW
20 Camsell River	Parts of 86 SE and 86 SW
21 Victoria–Vancouver	92 SE air and topo
22 Hudson Hope	94 SE exploratory topo only
23 Great Bear Lake	Parts of 96 B, C, F and G

Ground control points were still required, but because of the three cameras, one pointing directly downward, one pointing out to the right and one to the left, more ground could be covered on each pass, and as the newer aircraft could be flown higher, the number of control points needed was reduced.[3] The drawing of the charts was pushed ahead with wartime fervour. In July 1944, the final chart of the 221 sheet series was published.

The last sentence, although true, gives a false impression of the completeness of the series. On many of the 221 sheets there were large areas marked, 'unmapped territory'. The Trimetrogon photography had to be continued into the 1950s so that second editions of the more seriously deficient sheets could be prepared. Actually, a few small areas marked 'relief information incomplete' remained in the series until 1973 when information from the completed 1:250,000 series was used to clear up the deficiency.

During the Second World War the colour used for portraying flight information was changed from red to blue to conform with the American practice, and green and light green were used for the first two altitude tints above sea level. The contour intervals remain at 500 feet for most areas, but the altitude tints extend over 1000-foot intervals (i.e. two contour intervals). Red, being no longer required for the flying data, became the road colour. Black remains the colour for man-made features except the roads. Cities and towns are shown in yellow rather than the black cross-hatching of pre-war charts. Figure 19 shows the sheet layout, while Figure 20 is a section taken from a typical sheet in the series.

The 1:500,000 Topographic and Air Chart Series

In 1958 it was decided to change the scale of the series from 1:506,880 to 1:500,000. Two editions continued to be published, the topographic base and the aeronautical edition which consists of the same base maps with a blue air information overprint.

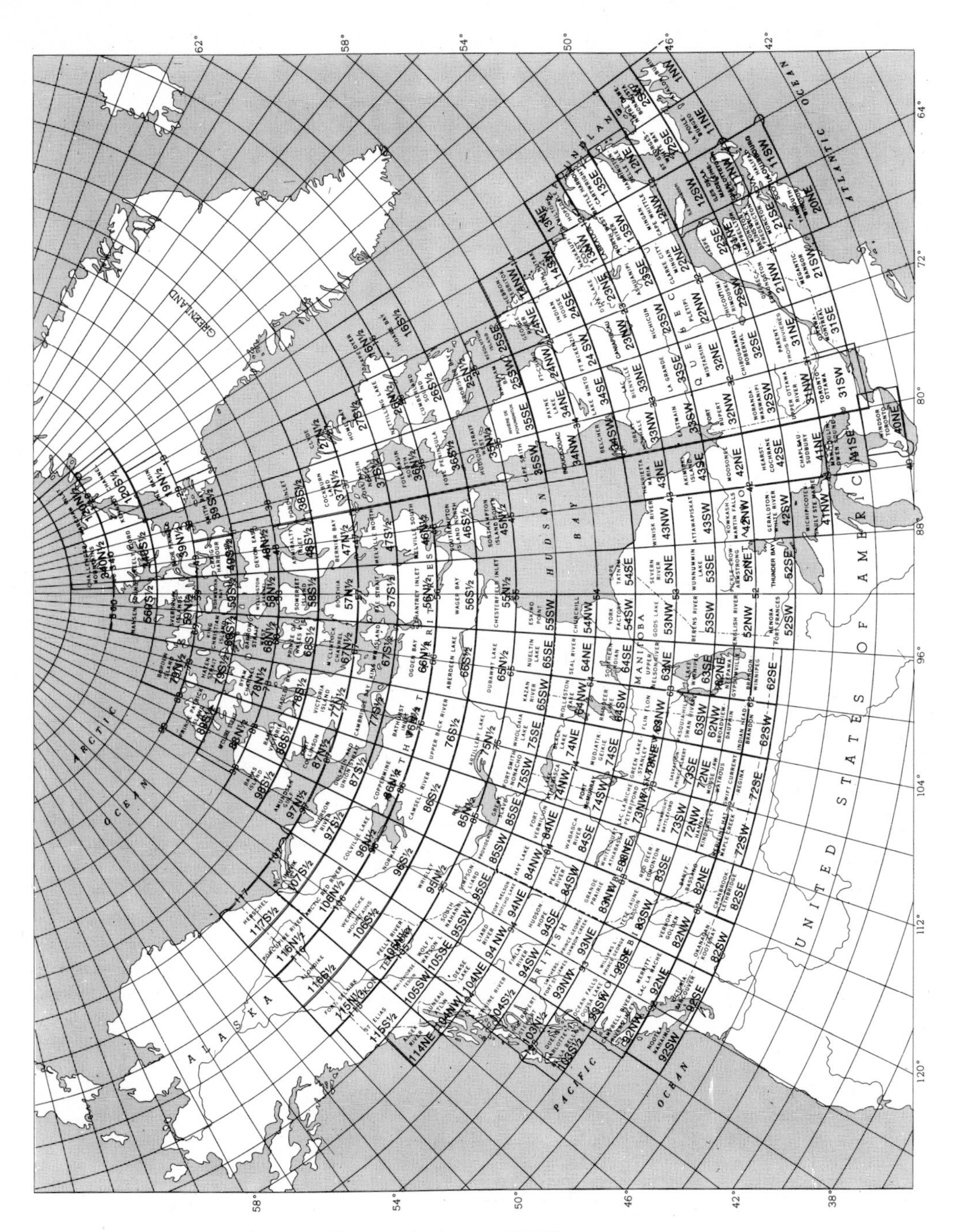

Figure 19 Index to 1:500,000 series

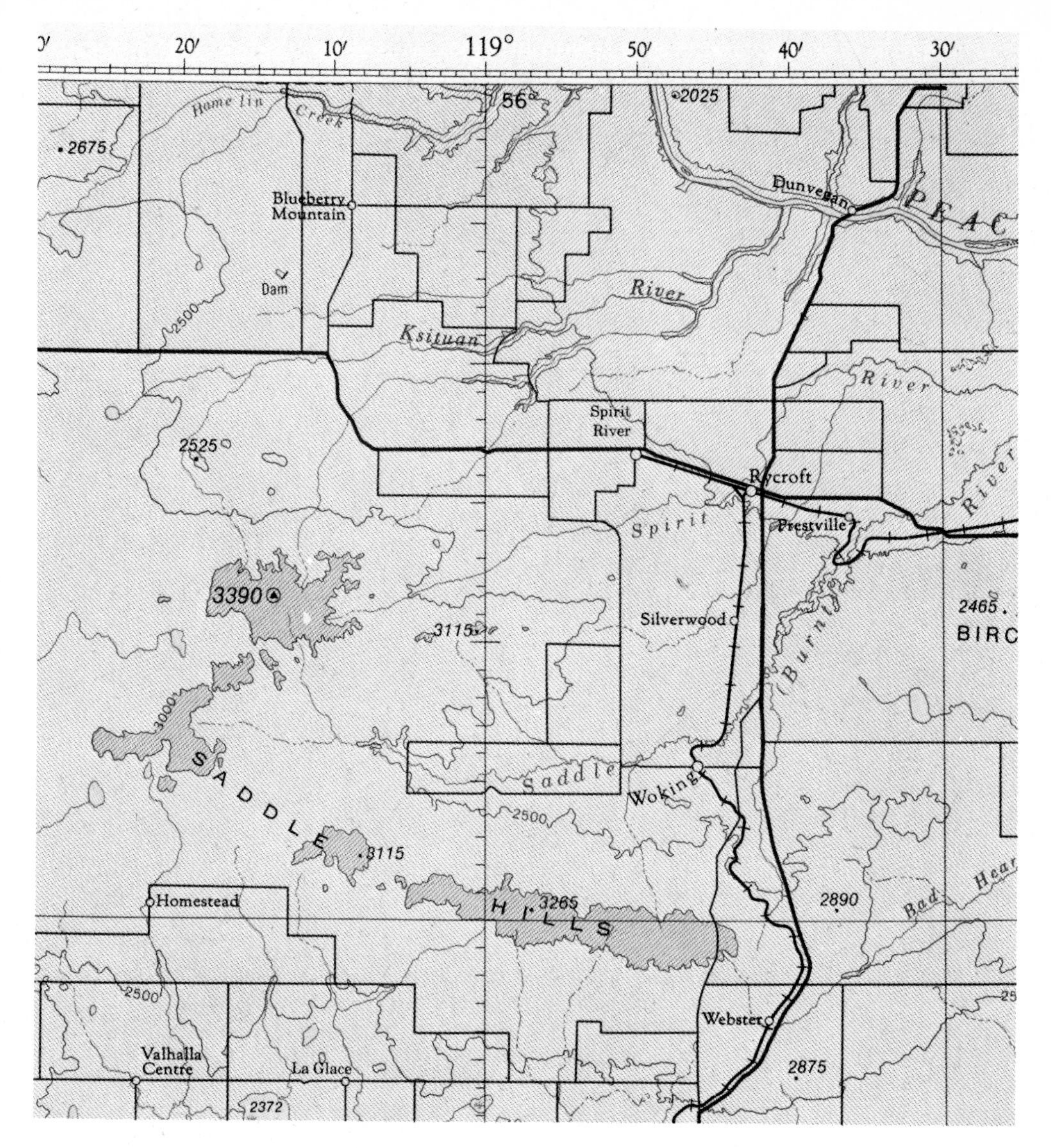

Figure 20 Section from a 1:500,000 sheet

During the early 1970s there was growing dissatisfaction among pilots with the format of the 1:500,000 series which had not changed since the early 1930s when planes had a much shorter flight radius and flew at much slower speeds. American charts at this scale, many of which showed large extents of Canadian territory, were printed on both sides of over-sized paper. Thus one American chart would cover four times the area of a Canadian chart, and both were sold at about the same price. In 1978 it was decided that Canadian charts should adopt a similar format. The first chart, Toronto, was published in September 1979. Each side of the sheet covers 2° of latitude and 8° of longitude, thus covering a quadrangle 4° by 8°. Fifty such charts will cover all of Canada when the series is completed.

The new series has other changes besides the enlarged format and back-to-back printing. Green is no longer used as an altitude tint, thus returning the height depiction to the original shades of brown used in the 1930s. Hill shading is used to show sudden changes in elevation. Yellow continues to be used for cities and towns, but now it is also used, in small yellow circles, to indicate villages. Roads, railways and power lines are symbolized, but all are given a somewhat subdued presentation by being printed in grey. Two colours are used for air information, blue for radio aids to navigation including flight paths, and magenta for airport information, critical elevations and obstructions. It is interesting to note the

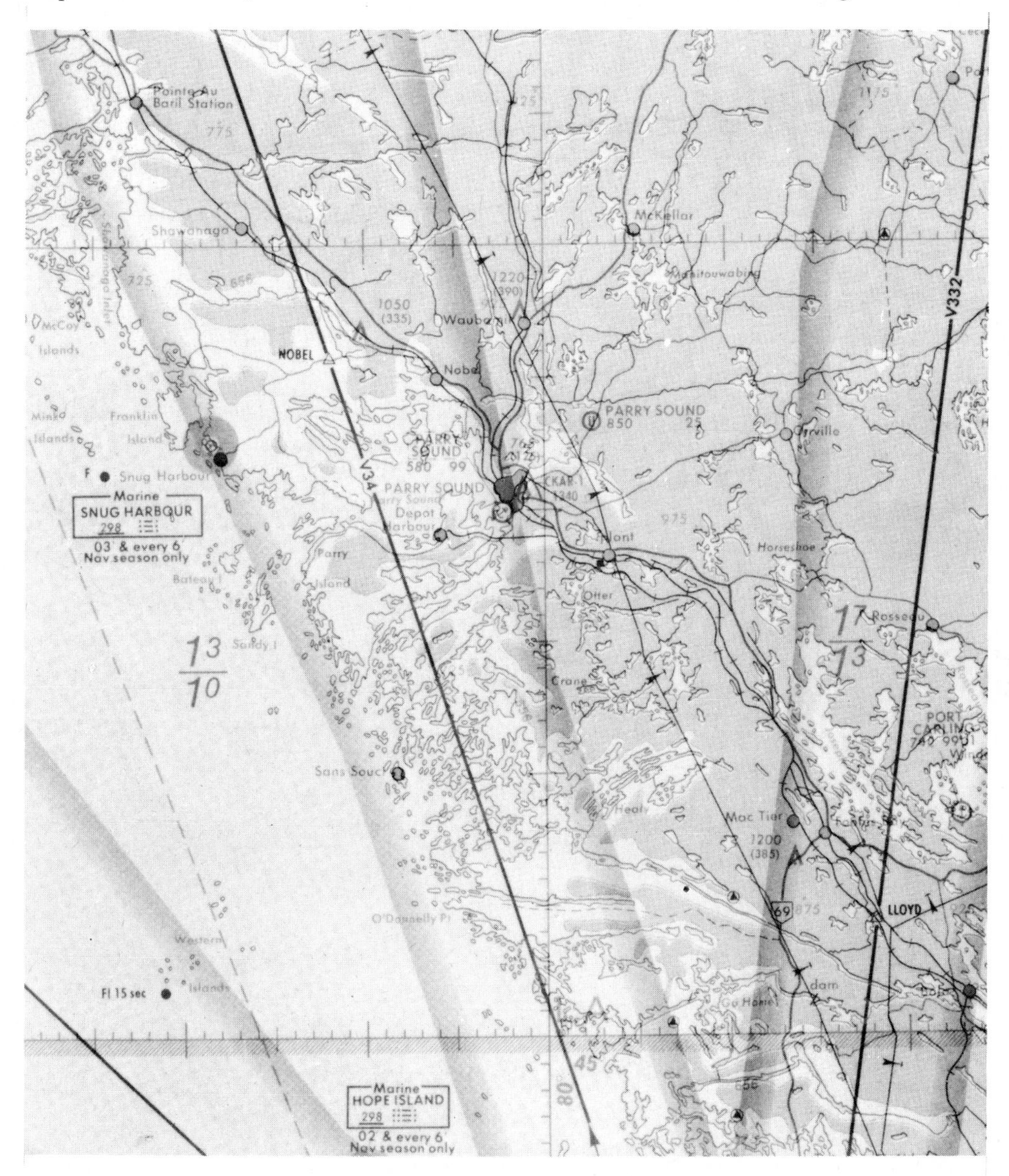

Figure 21 Section from a New Style 1:500,000 chart

prominence of the presentation of radio aids to navigation and the relegation to the background of features such as railways and power lines which were considered so important forty years ago. Figure 21 is a section taken from the Toronto sheet.

1:1,000,000 World Aeronautical Chart (WAC)

Canada's first attempt at a 1:1,000,000 aeronautical chart was made in 1945 when NTS sheet 83 – Athabaska River, was produced. This was a wartime experiment, and the post-war air industry did not seem to need a series at this scale. However, by 1948 there was a clear indication that a small-scale air chart series was needed. It was then decided to produce a series of aeronauntical charts at the scale of 1:1,000,000 drawn to the specifications set out by the International Civil Aviation Organization (ICAO), an agency of the United Nations based in Montreal. These charts would have as sheet lines the primary quadrangles of the NTS system, and thus the smallest scale of the National Topographic System finally came into being. The topographic data was derived from the 1:506,800 sheets. Man-made features, with the exception of roads, are shown in black; shorelines and drainage are shown in blue. Roads are in brown. Relief is given by contours and altitude tints. When the series was completed in 1961, 64 sheets covered all of Canada.

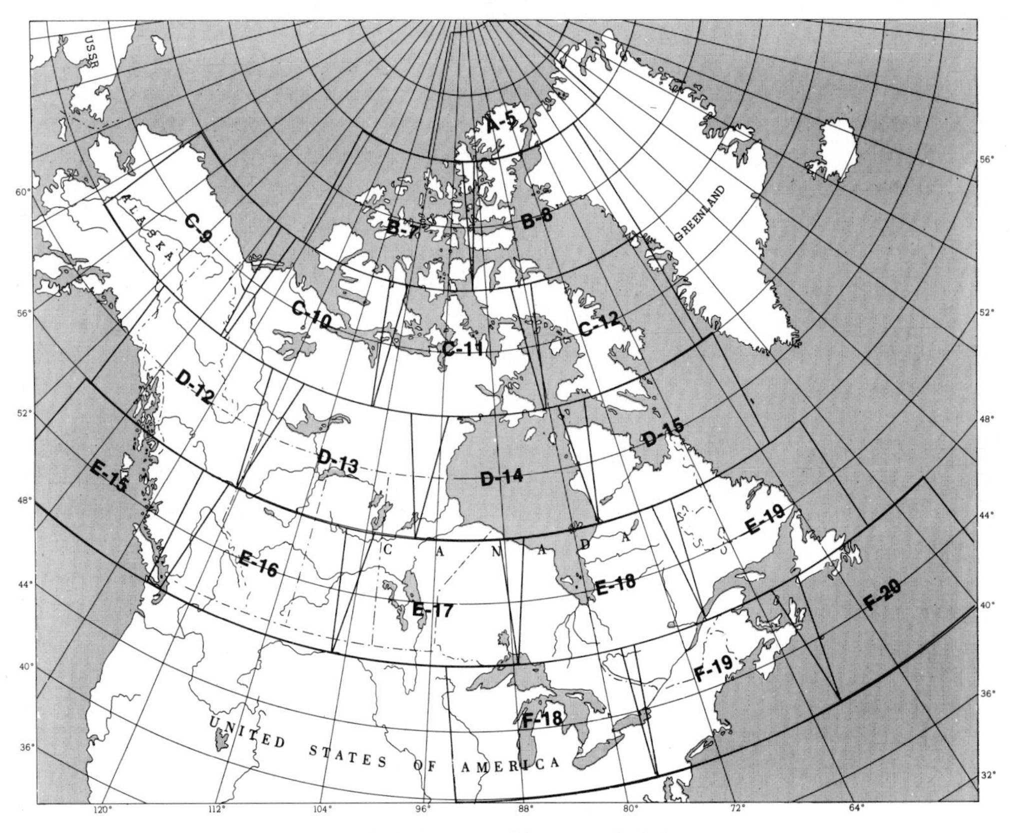

Figure 22 Index to world aeronautical charts

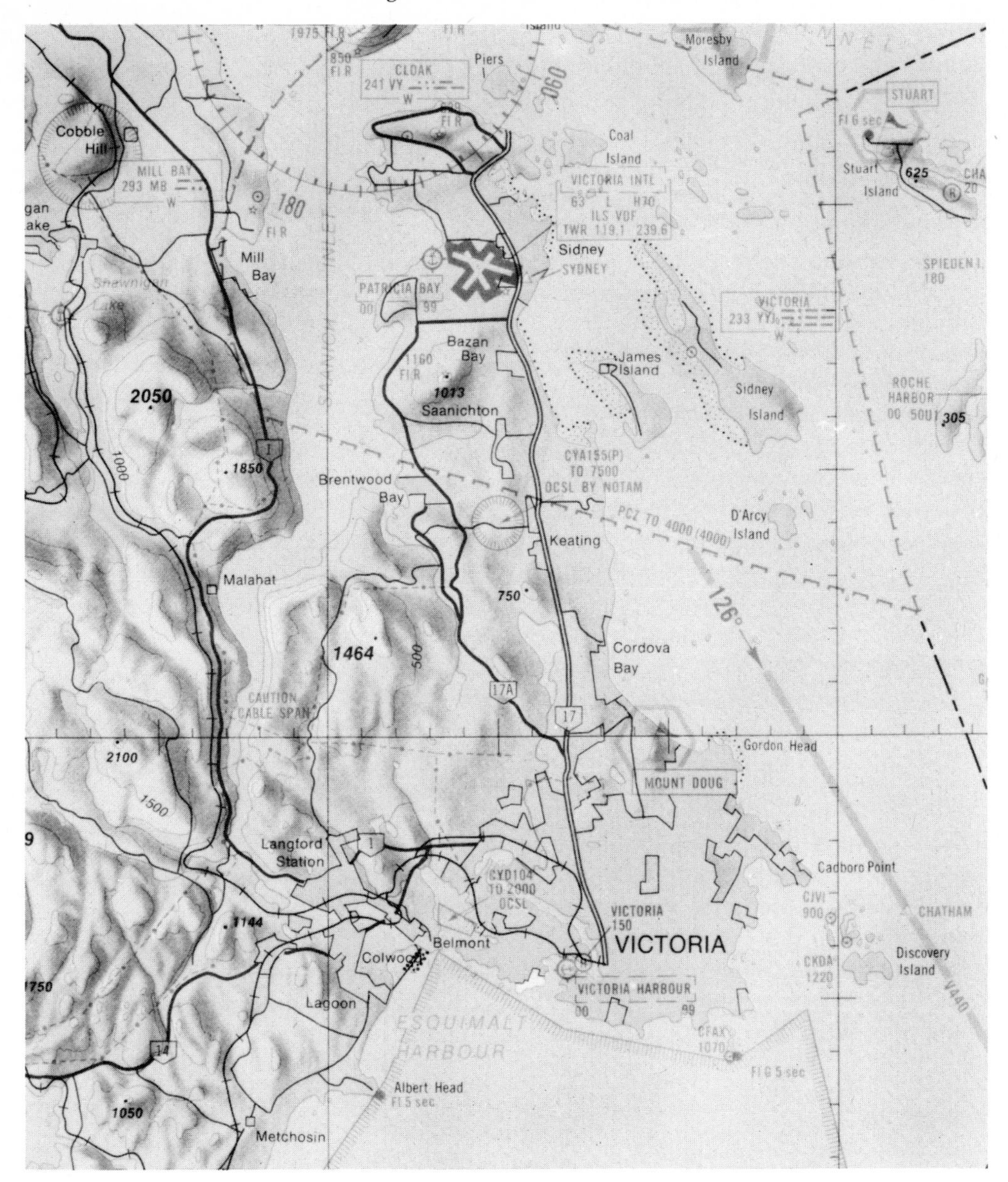

Figure 23 Section from a 1:250,000 VTA chart

In 1972 Canada followed the lead of other countries and modernized the presentation of the WAC. Larger paper was used, and printing was done on both sides. The sheet layout is as shown in Figure 22, and Canada is now covered by 19 sheets instead of the previous 64.

These new style sheets made use of hill shading as well as altitude tints to give a very graphic portrayal of the terrain. Roads are in brown, but other man-made detail is in black. Hydrography is in blue.

The topographic bases of the old style WAC charts are being kept in stock until the area depicted on them is mapped by sheets of the 1:1,000,000 IMW (Chapter 10 describes this series), at which time they are allowed to go out of stock. Thus, the old style WAC will disappear in 1980.

1:250,000 Visual Terminal Area Charts

The charts of this series, usually referred to as VTA charts, cover the major Canadian airports and will in time be extended to those of secondary importance. They are drawn with the airport near the centre of the sheet, and depict the terrain with contours, hypsometric tints at 1000-foot intervals and hill shading. The requirements of the aviator navigating by map reading are seen in the design of this series. Railroads are in black, roads in brown, and villages, towns and cities in bright yellow.

An interesting feature on charts of this series is the identification and marking of visual check points. These are distinctive map features that even the most indifferent map reader could not mistake. Points such as bridges over rivers, distinctive islands, and small towns are circled in purple and are named. Pilots flying into the airport are required to use these points for terse and unambiguous reporting of their position.

The following sheets in this series have been published: Vancouver, Winnipeg, Toronto and Montreal. Sheets for Edmonton and Calgary are in preparation. Figure 23 is a section taken from the Vancouver VTA chart.

Military Air Charts

The remaining types of aeronautical charts to be described are those produced for use by the Canadian armed forces and are available to the public only on the understanding that they will NOT be used for air navigation. The reason for this is quite simple. These charts are designed for specially trained military pilots, and they do not carry the usual civilian aids to navigation. However, they do give a graphic portrayal of the terrain that is not available in any civilian map series, and thus have uses that are not in any way connected with flying, such as geographical studies, education, ecological investigations, tourism, etc.

Joint Operations Graphics (Air)

These sheets, which are published at 1:250,000 scale, are designed for military operations in which there is close contact and continued communication between the air crews and the forces on the ground. By employing tints of green and buff, and hill shading, a very good picture of the ground is given. Forested areas are shown by a green stipple and care is taken in the cartography not to let the forest depiction confuse the hill shading. Roads are shown by parallel black lines with solid or broken brown fill distinguishing the number of lanes and the type of road construction (all-weather hard surface, all-weather loose surface, or dry weather). Other man-made detail is in black. Drainage is in blue with small lakes in medium blue and large water bodies in light blue. Cities and towns are outlined and filled with a screened brown which produces a salmon pink tint. Air information, including airport beacons, radio facilities and isogonic lines, critical altitudes are printed in a mauve colour officially known as purple-blue-purple.

The sheets of this series use the IMW 1:250,000 sheet lines and thus are 1° of latitude and 2° of longitude in extent. They are numbered in accordance with the IMW system for 1:250,000 sheets. The sheets available are listed in Table 15. Figure 24 is a section taken from a typical Canadian JOG (Air) sheet.

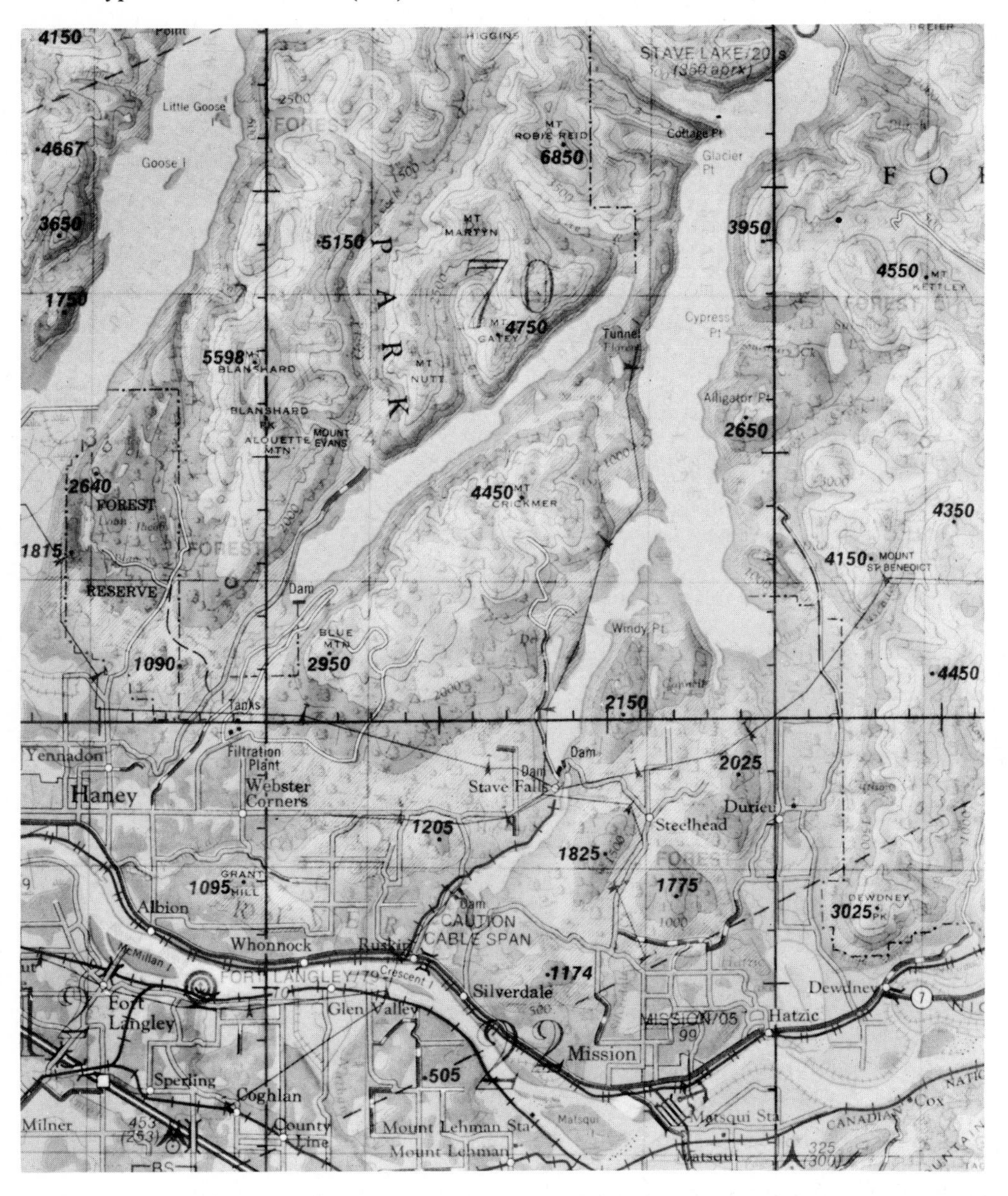

Figure 24 Section from a JOG (Air) chart

Table 15 Sheets Available in the Joint Operations Graphic Series

NK 17	sheets	2, 3, 4, 5, 6
NL 17	sheets	3, 6, 9, 12
NL 18	sheets	1 to 11
NL 19	sheets	1, 2, 3, 4, 5, 6, 7, 9, 12
NL 20	sheets	1, 4, 7, 10
NM 9	sheets	3, 5, 6, 9
NM 10	sheets	4, 7, 8, 10, 11
NM 12	sheets	5
MM 14	sheets	8, 9
NM 17	sheets	3, 6, 9, 12
NM 18	sheets	1 to 12
NM 19	sheets	1 to 12
NM 20	sheets	1, 4, 7, 10
NN 9	sheets	10, 11
NN 12	sheets	5, 6, 11
NO 15	sheets	1, 4, 5
NP 9 and 10	sheets	8, 12
NP 15 and 16	sheets	1, 2, 5, 6, 9, 13
NR 7 and 8	sheets	8, 9
NR 9 and 10	sheets	8

Note: Correct as of Feb. 1979.

Joint Arctic Weather Station Charts

This is a small series of five charts covering the approaches to Mould Bay, Isachsen, Cornwallis Island, Eureka and Alert. They are drawn to specifications similar to those of the JOG (Air) series and are at the same scale, namely, 1:250,000. These charts are designed to aid in visual or radar approaches to these weather stations where climatic conditions often make landings difficult.

Station	*Latitude*	*Longitude*
Alert	82°	64°
Cornwallis Island	75°	96°
Eureka	80°	86°
Isachsen	79°	104°
Mould Bay	76°	120°

Low Level Pilotage Charts

The charts of this 1:500,000 series began to appear in 1962 when they were produced to cover certain RCAF bases where they were used for the training of pilots in high speed, low level operations. In the design, emphasis was given to ground features that could be identified at a glance by the pilot as they passed beneath his plane. Shaded relief was used to give good ground portrayal and to provide ground recognition when radar was used in navigation. Altitude tints were employed to accentuate changes in the elevation of the ground. Roads, railways, towns and villages are shown in detail appropriate to the chart scale. Air information is shown in blue, and includes airdromes by symbol or runway pattern, selected radio aids, air defence identification zones, critical elevations and isogriv lines.

The future of this series is in doubt. Some believe that the new style civilian 1:500,000 pilotage charts, described previously, will replace it. The extent of the present series is shown in Table 16A

Table 16A Low Level Pilotage Charts —Edition Data

NTS number	Base edition year	Air information year
11NW	3–73	Apr 76
11SW	4–77	Sep 77
20NE N½ 21SE S½	3–73	Aug 73
21NE	5–76	Jun 77
21NE S½ 21SE N½	4–77	Jul 77
22SE	3–73	Jun 76
52NW	3–77	Jan 78
52SW	2–73	Apr 74
53SW	2–74	Apr 76
62NW	4–77	Dec 77
62NE	3–77	Mar 78
62NE S½ 63SE N½	3–77	Feb 78
63NW	2–73	Nov 76
63SW	3–77	Dec 77
63SE	2–74	Nov 76
64NW	3–75	Jan 76
64SW	2–67	Jun 73
72NW	4–73	Nov 76
72NE	3–74	Mar 74
73NW	3–72	Jul 76
73NE	3–73	Mar 74
73SW	4–73	Feb 74
73SE	4–76	Apr 77
74NW	3–73	Apr 74
74NE	2–64	Mar 76
74SW	5–73	Mar 74
74SE	2–67	Jun 73
82NE	3–73	May 76
83NW	4–76	Jul 76
83NE	4–73	Nov 77
83SW	5–74	Jan 76
83SE	4–76	Jun 77
84NE	3–73	Jan 77
84SE	4–73	Feb 74
84SW	2–73	Jul 76

Note: The location of these charts may be found in Figure 19.

Low Level Aeronautical Charts

This is a small series of 16 charts drawn at 1:250,000 scale. In appearance they resemble the low level pilotage charts, just described, and they were produced in 1963 and 1964 for low level flight training. They are being replaced by the JOG (Air) series. Table 16B lists the charts that are available.

Table 16B Low Level Aeronautical Chart 1:250,000—Edition Data

NTS number	Base edition year	Air information year
73E	1–64	Mar 64
73F	1–64	Apr 76
73G	1–64	May 76
73J	1–63	May 76
73K	1–63	May 73
73L	1–63	May 73
73M	1–64	Jun 75
73N	1–64	May 73
73O	1–63	Oct 63
74B	1–63	Oct 63
74C	1–63	May 69
74D	1–64	May 69
83H	1–64	Sep 64
83I	1–63	May 76
83P	1–64	Feb 64
84A	1–63	Apr 69

9. The 1:25,000 Series

Canada's 1:25,000 maps were originally produced solely for military use. The origins of the series can be traced back to the maps of military training areas produced during the First World War. By 1916 military operations in France had solidified into trench warfare in which both offensive and defensive operations were based on massive artillery barrages. The aiming of the hundreds of guns that were used in such tactics required detailed maps of

Table 17 Large-Scale Military Maps Produced Between 1916 and 1918

Survey division number	Name	Scale	Date
104	Petawawa (4 sheets 57NW, 57NE, 57SW, 57SE*), Ont.	1:20,000	Undated (magnetic declination given for 1911)
112	Valcartier, Quebec (90NW)	1:20,000	Undated, probably 1916
120	Halifax, NS	1:20,000	1917
117	Niagara and vicinity, Ont.	2 inches to 1 mile (1:31,680)	1916
121	Calgary, Alta.	3 inches to 1 mile (1:21,120)	1916
113	Camp Borden, Ont.	2 inches to 1 mile	1916
115	Camp Borden, Ont.	3 inches to 1 mile	1917
131	Kingston and vicinity, Ont.	2 inches to 1 mile	1917
132	Manoeuvre area around Toronto, Ont.	2 inches to 1 mile	1917
133	Halifax (3 sheets 133a, 133b, 133e), NS	2 inches to 1 mile	1917
119	Sarcee Reserve and Camp, Alta.	1:20,000	1916

* Note: 57 refers to the 1:63,360 sheet that was quartered to make the 1:20,000 sheets.

the battlefield, and these were produced by the mapping agencies of both the French and British armies. Information about these military arts was sent back to Canada so that reinforcements could be trained in the appropriate battle skills. As the battlefield maps being used in France were at the scale of 1:20,000, maps at or near this scale were prepared for a number of Canadian training centres. Table 17 lists the maps produced between 1916 and 1918. It will be noted that the scales ranged from 1:20,000 to 1:31,680 but the detail on the maps was almost exactly the same as that shown on the one-inch maps. In fact, in areas where the one-inch mapping preceded the larger scale, the one-inch sheets were simply enlarged and published with no additional detail. Where one-inch maps did not exist, the topographers were ordered to follow the one-inch mapping specifications without change.

After the war the military authorities continued to publish new editions of the training centre maps but made only modest extensions of large-scale topographic mapping into new areas. Canada was just too large, and too unmapped, to even consider extensive mapping at any scale larger than the one-inch. Certainly military planners would have liked more detailed maps, but they had to face reality. In a report made to the Board of

Table 18 Large-Scale Military Maps Published Between 1919 and 1939

GSGS number	Name	Province	Scale	Date	Remarks
145	Sydney	NS	1:31,680	1919	145a is NE sheet 145b is NW sheet
178	Camp Hughes	Manitoba	1:20,000	1921	(Now part of Camp Shilo) In four sheets NW, NE, SW, SE
214	Camp Sarcee	Alberta	1:20,000	1924	
238	Petawawa	Ontario	1:50,000	1924	
241a	Victoria and vicinity	BC	1:40,000	1925	241a is north sheet, 241b is south sheet
284	Kingston	Ontario	1:31,680	1927	Revision of 131 of 1917. Reprinted with corrections 1934
299	Valcartier	Quebec	1:25,000	1931	
330	Comox	BC	1:25,000	1931	
331	Dundurn	Saskatchewan	1:20,000	1931	Four sheets NW, NE, SW, SE
379	Lonsdale	Ontario	1:25,000	1933	
381	Lac Du Bonnet	Manitoba	1:20,000	1932	RCAF Base
385	Halifax	NS	1:25,000	1934	
387	Winnipeg	Manitoba	1:25,000	1933	Numbered 62 H/14 SE
409	Thurlow Island	BC	1:25,000	1933	
415	Victoria	BC	1:25,000	1933	Eight sheets a to h
429	Masson	Quebec	1:25,000	1934	Rapid mapping exercise
432	North Gower	Ontario	1:25,000	1934	
482	Rimouski	Quebec	1:25,000	1938	
485	Chicoutimi	Quebec	1:25,000	1938	
504	Sussex	NB	1:25,000	1937	
505	Pointe aux Trembles	Quebec	1:25,000	1937	
506	St Johns	Quebec	1:25,000	1937	
507	Niagara	Ontario	1:25,000	1937	
508	Shilo	Manitoba	1:25,000	1937	Two sheets—north and south
509	Dundurn	Saskatchewan	1:25,000	1937	No. 331 reduced
525	Pine Ridge	Manitoba	1:25,000	1938	
548	Gloucester	Ontario	1:25,000	1939	

Topographic Surveys and Maps in 1925, Colonel J. S. Brown of the Canadian General Staff made the following statement:

> The ideal map for tactical operations is one on a scale of 1:20,000 with a 10′ VI. Such maps may be produced for small areas, but it is an impossibility to have maps on such a scale for extended areas, therefore, it has been accepted by the War Office and by the British Dominions, generally, that the most suitable map likely to be obtained is on the scale of 1 inch to 1 mile contoured at a Vertical Interval of 25 feet.[1]

During the period between the two world wars the Geographical Section of the General Staff had an establishment of about 30 military and 10 civilians. On an average, five one-inch quadrangles were published annually and still resources were found to publish training area maps at larger scales. Table 18 lists the sheets published between 1919 and 1939 at scales between 1:20,000 and 1:31,680.

In the production of these maps the specifications for the surveying and cartography continued to be virtually the same as those used for the one-inch mapping. All military topographic mapping was done on the plane table, and the drawing scale of 1:31,680 was used for all scales. If one-inch mapping was required, the plane table sheets were reduced to this scale; if 1:31,680 or larger scales were ordered, the plane table sheets were left at scale or enlarged.

During the 1920s there appear to have been some doubts among Canadian military authorities about the best scale for large-scale tactical maps. It would appear that the artillerists pressed for the 1:20,000 scale, because the artillery training centres of Camp Hughes, Sarcee, Dundurn and Petawawa were mapped at this scale. The infantry and cavalry seem to have preferred the 1:31,680 scale. This uncertainty was resolved abruptly and from afar. In 1932 the War Office in London decided that 1:25,000 maps would be standard for large-scale military maps throughout the British Army. The Canadian General Staff conformed to this decision, and from that year to the present the 1:25,000 scale has been standard for large-scale federal topographic mapping. In fact the first serious challenge to this decision did not appear until 1970 when the Province of Quebec began to produce the 1:20,000 topographic maps (See Chapter 13).

The years of the Second World War were years of great activity for the GSGS, but its work was for the most part at scales smaller than 1:25,000. Nevertheless, resources were found for the mapping of the new training centres required by the general mobilization. These are listed in Table 19.

The period following the Second World War in many ways followed the pattern set in 1919. Training camps were reduced in size or, in many cases, closed completely. The military mapping agency (by this time called the Army Survey Establishment) was instructed to concentrate on one-inch mapping with only a small proportion of its effort being directed to 1:25,000 mapping. Some training camp maps were up-dated and there was a small but interesting new use for large-scale topographic maps. The strategic Arctic settlements, Aklavik, Fort Resolution and Hay River, were mapped at 1:25,000.

A far more important use of 1:25,000 maps resulted from a new military training policy which was formulated in the Department of National Defence in 1952. It was argued that the modern tactics used in mechanized warfare required much larger training areas than those previously used by the Canadian Army. The first of these large areas to be developed was Camp Gagetown in New Brunswick. About 800 square miles of country consisting of forest, rough pasture and small farms, was expropriated. As the artillery training require-

Table 19 1:25,000 Military Maps Published Between 1940 and 1945

GSGS number	Name	Province	Date	Remarks
580	Tracadie	NS	1940	
589	Nanaimo	BC	1940	
601	Sydney	NS	1940	Derived from no. 145
602	Glace Bay	NS	1941	
628A	Kitchener	Ontario	1942	
628B	Ipperwash	Ontario	1942	Training centre
628C	Chatham	Ontario	1942	
628D	Guelph	Ontario	1942	
628	Meaford	Ontario	1942	Tank range
631A	Brandon	Manitoba	1942	Training centre
631B	Portage La Prairie	Manitoba	1942	Training centre
631C	Wetaskiwin	Saskatchewan	1942	Training centre
631D	Camrose	Alberta	1942	Training centre
631F	Prince Albert	Saskatchewan	1942	Training centre
631G	North Bay	Ontario	1942	Training centre
	In addition to the above, most sheets previously published in the older scales were brought out in 1:25,000 editions. The following are examples:			
115	Camp Borden	Ontario	1941	Derived from 1916 edition
130	Aldershot	NS	1941	Derived from 1917 edition
284	Kingston	Ontario	1941	

ment still demanded 1:25,000 maps, such maps had to be produced for this new camp. But the area involved could not be drawn on a single camp map; in fact 12 sheets were required for the complete coverage of the camp. It was decided therefore to extend the National Topographic System of sheet lines down to the 1:25,000 scale. At the time the 1:50,000 maps at the latitude of Camp Gagetown were published in half-sheets, each covering 15 minutes of latitude and 15 minutes of longitude. These were quartered to produce 7½ minute 1:25,000 sheets. The eight sheets within a complete 1:50,000 quadrangle were lettered a to h as illustrated in the diagram below[2].

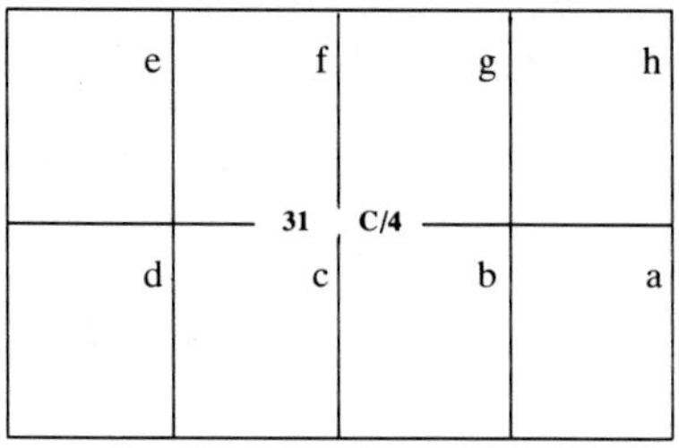

e	f	g	h
	31	C/4	
d	c	b	a

The specifications for this new 1:25,000 series were in most respects quite similar to those in use for the 1:50,000 scale. To avoid useless repetition of the material already covered in Chapter 5, only the differences between the 1:25,000 specifications and those for 1:50,000 mapping are described in the following paragraphs.[3]

MAN-MADE FEATURES. Smaller buildings, such as sheds and garages over 20′ × 20′ in size, are shown. More public buildings are labelled. Fences and walls of permanent construction are shown. City subways (underground railways) are shown as railways where they can be seen from the air, and as tunnels where they are underground. (On 1:50,000 maps, only the open portion of the subway is depicted.)

RELIEF. As is to be expected, a closer contour interval is used. In 1978 all existing 1:25,000 sheets have either a 10-foot or 25-foot interval, except two sheets published in 1976 covering Sherbrooke, Quebec. These have a 5-metre interval. When new sheets are published they will have metric contours at 2.5, 5 or 10-metre intervals depending on the relief.

ROADS. Cased roads (i.e. road symbols consisting of two parallel black lines filled with a solid or broken red line) continue to be used on 1:25,000 maps despite the fact that they have been abandoned for the 1:50,000 series. As the complete hierarchy of road classifications, giving the number and width of lanes and the type of road, be it paved or gravel, can be shown by the red road fill, the additional colour (orange) used on the 1:50,000 series is not required for 1:25,000 maps.

VEGETATION. The older sheets of the 1:25,000 series (i.e. those published before 1964) showed, by different symbols, coniferous and deciduous forests. This distinction was dropped to reduce the amount of fieldwork.[4]

At the 1:25,000 scale, landmark trees and well-established hedges that are over 1000 feet in length are shown by green dots 0.65 mm in diameter. Vineyards, orchards and tree nurseries are depicted in the same manner as on the 1:50,000 series but smaller installations can be shown (see Appendix 2 for minimum sizes of features).

HYDROGRAPHY. The only difference in the depiction of the hydrography is that smaller features may be shown. The minimum sizes of drainage features are included in Appendix 2.

Table 20 lists the special military maps that were available in 1956 when the first of the NTS 1:25,000 series became available. These older 1:25,000 maps all have a military significance in that they cover training areas or (in the north) areas considered to be of strategic importance at the time they were produced. They do not follow any uniform set of specifications but anyone familiar with Canada's one-inch or 1:50,000 symbolization would have no trouble in using them. As indicated in the appropriate column of Table 20, some carry the Modified British Grid. This grid system is explained in Chapter 15.

MILITARY TOWN PLANS. The 1:25,000 mapping of Camp Gagetown was followed by the mapping of Camp Valcartier in Quebec and Camp Wainwright in Alberta. This work was still going on in 1959 when the Canadian Government became concerned about the protection of Canadian cities against an atomic attack. There was an immediate requirement for 1:25,000 coverage of the 17 largest Canadian cities. These cities were Calgary, Edmonton, Halifax, Hamilton, London, Kitchener–Waterloo (one sheet), Montreal (two sheets), Niagara Falls, Ottawa, Quebec, St John, New Brunswick, St John's, Newfoundland, Toronto (two sheets), Vancouver, Victoria, Windsor and Winnipeg.

The initial demand was for a special map centered on the city and extending outwards to cover the immediate suburbs and the city's airport. (These were originally called military town plans but in 1969 the series was renamed military city maps.) This special map series originally included in its design all features normally shown on a standard 1:25,000 map

Table 20 1:25,000 Military Maps in Stock 1956

Sheet name	Sheet lines latitude and longitude	Date	Contour interval	Grid	Remarks
Aldershot	45°04–45°09 64°26–64°35	1952	20′	UTM	5 colours
Aldershot	45°00–45°07 64°22–64°37	1956	25′	UTM	5 colours plus purple grid
Aklavik	68°10–68°17 134°54–135°10	1951	No contours	UTM	Black, blue and green
Bedford Basin	44°40–44°47 63°30–63°44	1955	25′	UTM	3 colours
Barriefield	44°13–44°20 76°10–76°28	1953	25′	UTM	3 colours
Brewers Mills	44°20–44°27 76°12–76°30	1953	25′	UTM	3 colours
Borden	44°14–44°19 79°53–79°57	1945	20′	Modified British	4 colours
Borden	44°07–44°22 79°45–80°00	1956	10′	UTM	5 colours
Chilliwack	49°00–49°11 121°52–122°07	1955	50′	UTM	10 sheets 5 colours
Chicoutimi	48°23–48°27 70°57–71°04	1937–43	25′	Modified British	3 colours partly contoured
Connaught Range	45°18–45°24 75°57–71°04	1926–35	12.5′	Modified British	4 colours
Debert	44°21–44°29 63°20–63°32	1942	25′	Modified British	3 colours
Dundurn	51°45–51°55 106°31–106°44	1948	25′	UTM	4 colours
Dundurn	51°45–52°00 106°30–107°00	1956	10′	UTM	5 colours
Farnham	45°15–45°22 72°52–73°07	1956	10′	UTM	5 colours
Fort Osborne	49°45–49°52 97°06–97°22	1941	25′	Modified British	3 colours
Fort Resolution	61°06–61°15 113°31–113°46	1952	25′	UTM	4 colours
Glace Bay	46°08–46°15 59°51–60°05	1942	25′	Modified British	3 colours
Goldstream	48°25–48°30 123°30–123°45	1939	25′	Modified British	3 colours
Halifax	44°34–44°40 63°30–63°44	1955	25′	UTM	4 colours
Hay River	60°45–60°53 115°36–115°51	1952	25′	UTM	4 colours
Ipperwash	43°05–43°14 81°50–82°02	1956	10′	UTM	5 colours
Kingston	44°13–44°20 76°28–76°45	1953	25′	UTM	3 colours
Lawrencetown	44°34–44°40 63°16–63°30	1955	25′	UTM	5 colours
Lonsdale	44°15–44°22 77°00–77°15	1933	25′	Modified British	3 colours
Malahat	48°30–48°37 123°30–123°45	1939	25′	Modified British	3 colours
Meaford	44°38–44°44 80°35–80°47	1955	25′	UTM	5 colours
Meaford	44°37–44°45 80°37–80°52	1956	25′	UTM	5 colours

Sheet name	Sheet lines latitude and longitude	Date	Contour interval	Grid	Remarks
Metchosin	48°17–48°25 123°30–123°45	1939	25′	Modified British	3 colours
Niagara-on-the-Lake	43°10–43°16 79°03–79°13	1956	25′	UTM	3 colours
Nanaimo	49°06–49°15 123°54–124°05	1941	25′	Modified British	4 colours
Porter Lake	44°40–44°47 63°10–63°25	1955	25′	UTM	4 colours
Penobsquis	45°42–45°50 65°10–65°24	1942	50′	Modified British	Black and white
Petawawa	45°52–46°02 77°15–77°26	1953	20′	UTM	4 colours
Petawawa	44°52–46°02 77°15–77°37	1956	25′	UTM	5 colours
Pine Ridge	49°56–50°03 96°44–96°57	1938	12.5′	Modified British	3 colours
Pointe aux Trembles	45°36–45°44 73°24–73°38	1937	25′	Modified British	3 colours
Prince George	53°51–53°47 122°35–122°50	n.d.	50′	Modified British	4 colours
Rimouski	48°25–48°28 68°30–68°35	1937–44	25′	Modified British	3 colours partly contoured
Saanich	48°30–48°37 123°15–123°30	1939	25′	Modified British	3 colours
Saint John NB	45°11–45°18 65°55–66°10	n.d.	25′	Modified British	3 colours
St John's Quebec	45°16–45°24 73°14–73°26	1937	25′	Modified British	3 colours
Sambro	44°26–44°33 63°30–63°44	1955	25′	UTM	4 colours
Sarcee	50°56–51°02 114°06–114°18	1950	20′	Modified British	4 colours
Sarcee	50°55–51°02 114°05–114°22	1955	25′	UTM	5 colours
Shawnigan	48°37–48°45 123°30–123°45	1939	25′	Modified British	3 colours
Shilo	49°30–50°00 99°00–99°45	1954	10′	UTM	4 colours 4 sheets
Sidney	48°37–48°45 123°15–123°30	1935	25′	Modified British	3 colours
Sooke Bay	48°20–48°25 123°45–124°00	1939	25′	Modified British	3 colours
Sussex	45°40–45°47 65°24–65°36	1941	25′	Modified British	3 colours
Sussex	45°37–45°45 65°22–65°37	1954	25′	UTM	5 colours
Sydney	46°08–46°15 60°06–60°19	1941	25′	Modified British	3 colours
Tracadie	47°25–47°31 64°58–65°13	1941	25′	Modified British	3 colours
East Thurlow Island	50°21–50°28 125°19–125°34	1934	100′	Modified British	4 colours
Valcartier	46°00–46°57 71°22–71°45	1939	25′	Modified British	3 colours
Valcartier	46°45–47°07 71°22–71°45	1954	25′	UTM	5 colours
Victoria	48°22–48°30 123°15–123°30	1939	25′	Modified British	3 colours

Table 20—*continued*

Sheet name	Sheet lines latitude and longitude	Date	Contour interval	Grid	Remarks
Wainwright	52°40–52°50 110°54–111°07	1953	25′	Modified British	4 colours 2 sheets
Wainwright	52°37–52°52 110°30–111°15	1956	10′	UTM	5 colours 2 sheets

The following colours were employed in military mapping before 1956:
3-colour maps—blue drainage, brown contours, black for all other detail including black tree symbols.
4-colour maps—blue drainage, brown contours, green vegetation, black for all other detail.
5-colour maps—same as for 4-colour maps plus red road fill.
Two types of grid were used before 1956, the Modified British Grid and the Universal Transverse Mercator.

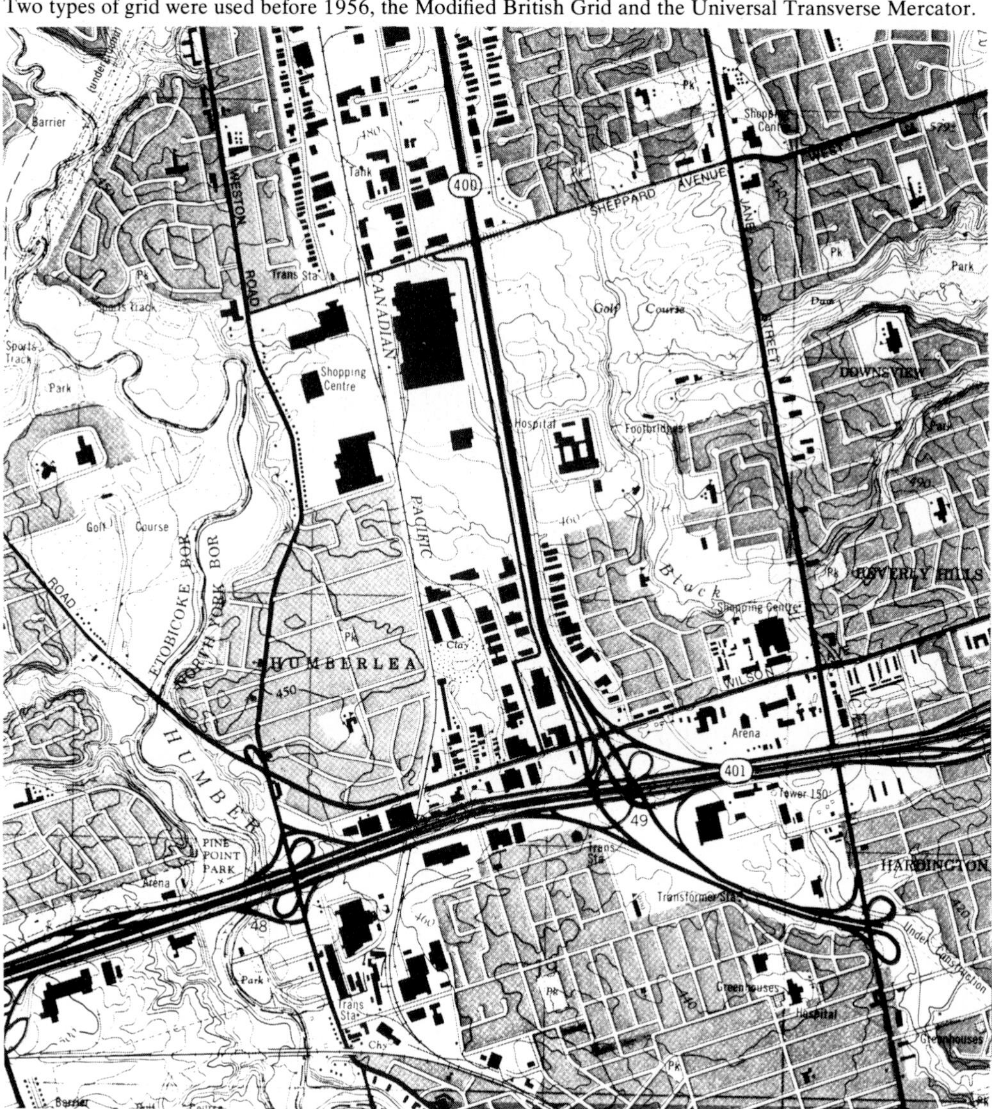

Figure 25 Section from a 1:25,000 NTS sheet

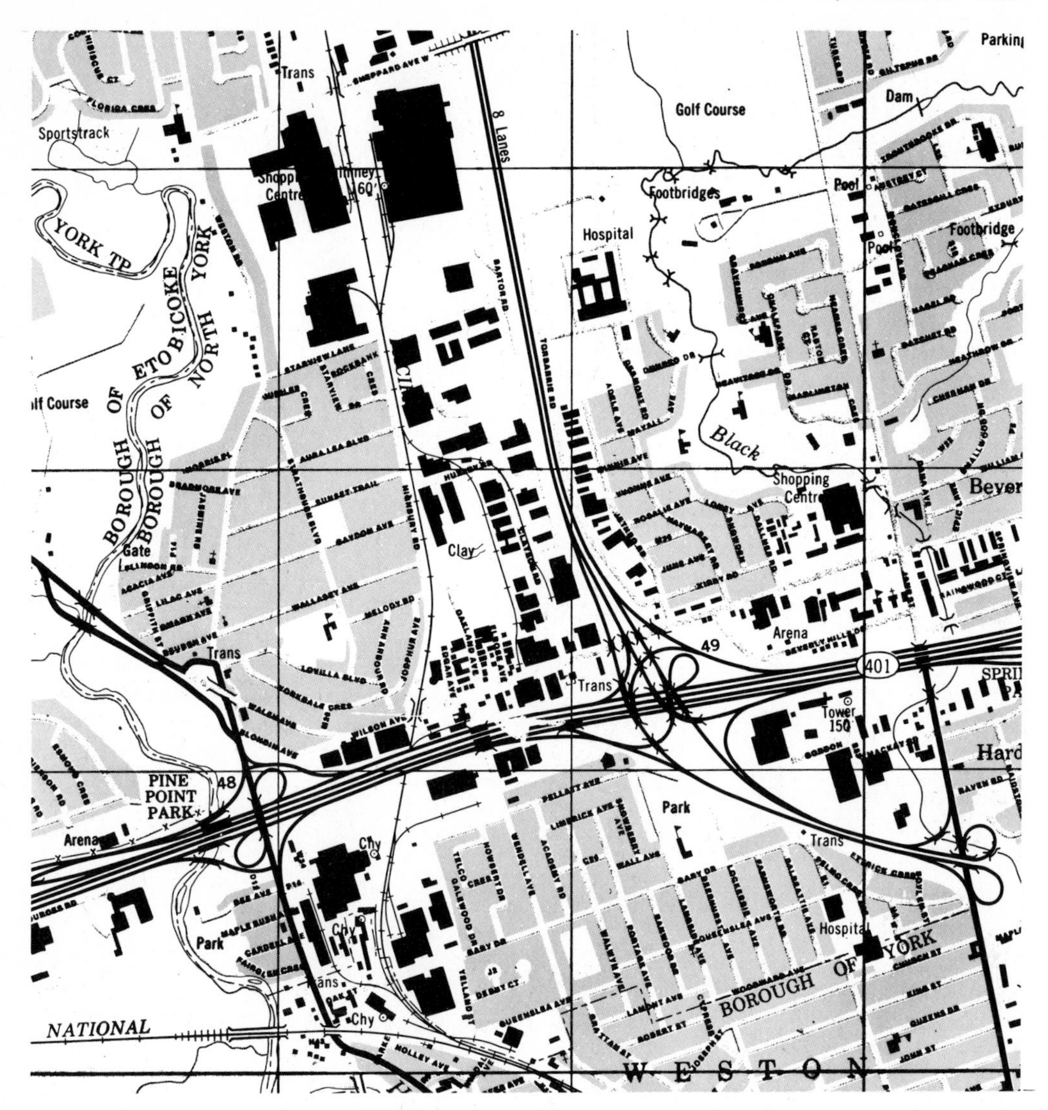

Figure 26 Section from a 1:25,000 military city map

plus a detailed identification of a number of installations of special military and civil defence importance such as hospitals, police stations, power stations, etc. All such features were tabulated in the margin together with UTM grid reference. A street index, also grid-referenced, was printed on the back of the map.

Military town plans were classified as 'Restricted' which meant that they could not be sold to the public. This restriction was placed on them not because any of the installations shown was a military secret, but because it was thought unwise to display the complete 'nerve system' of the city on an open document. But the surveying and cartography required to produce these military specials could be easily used to produce standard NTS 1:25,000 maps, and this was done using the specifications originally worked out for the

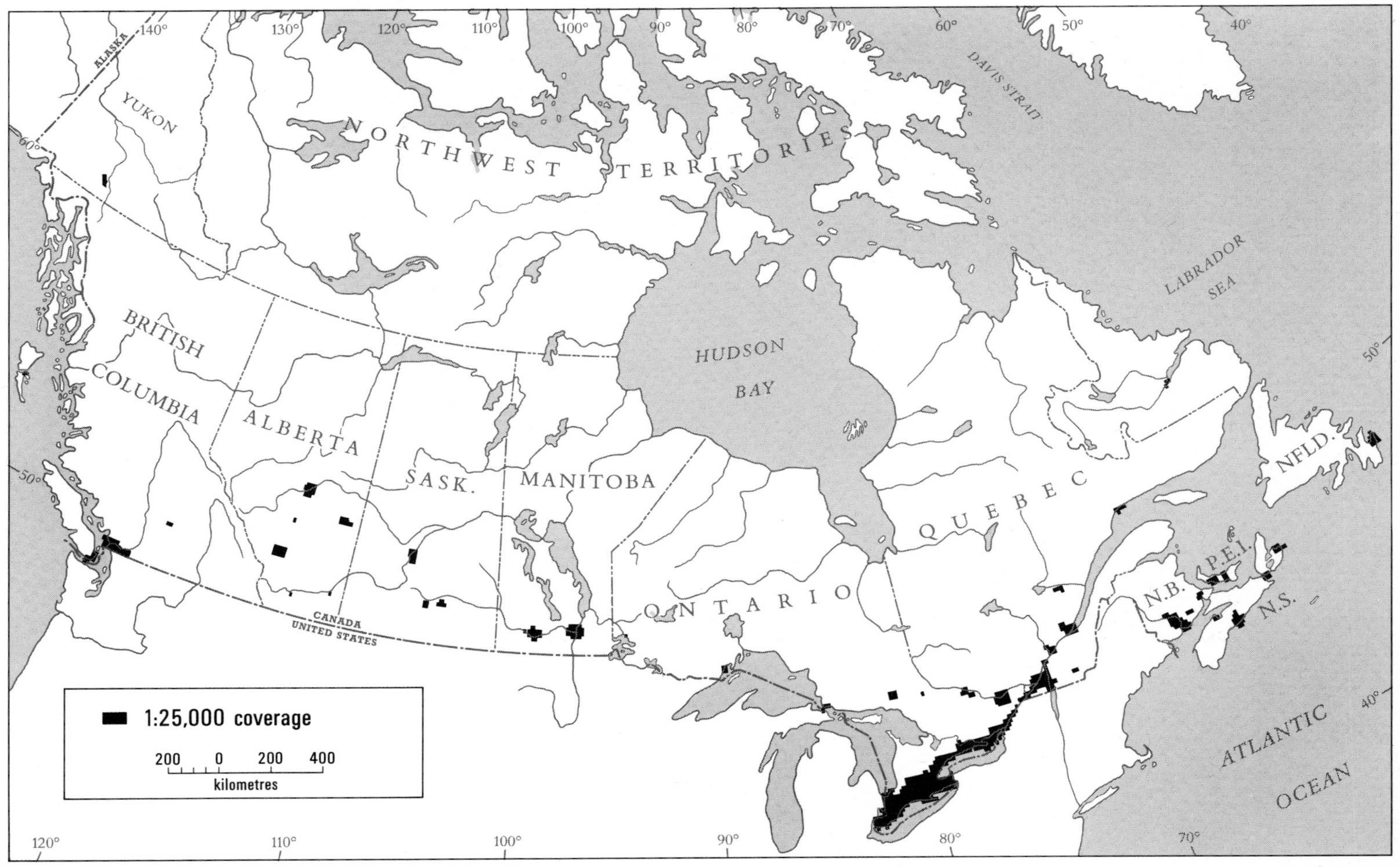

Figure 27 1:25,000 NTS coverage

Gagetown maps. Even with these maps available to the public there was a demand for a civilian version of the military city map which was centered on the city and displayed the whole city on one (or at most two) sheets of paper. Since 1970, an unrestricted version of the MCM has been available differing from the military version only in a reduction in the amount of labels used to identify public buildings.

In 1970 the military mapping authorities redesigned the military city maps to improve their legibility and to facilitate their revision. In the redesign the specifications of the MCM differed considerably from those used for NTS sheets at the same scale. Figure 25 shows an area of Toronto depicted by the standard NTS sheet and Figure 26 the same area on the MCM. Had these maps been in colour the difference between the two would have been more striking. The following are the main differences:

Streets are not cased in the military city maps. Through routes are shown in a gold-brown line 0.6 mm wide while streets other than through routes are in grey at 0.4 mm in width.

Built-up areas are depicted in buff (screened gold-brown) rather than the standard pink (screened red) of the NTS sheets. The buff shows the built-up areas quite well, and at the same time accentuates the grey street pattern.

Green is used to depict vegetation, but in these sheets a special military significance is given to the colour. Solid green is used to depict parkland which is suitable for bivouacs, while pale green (screened green) is used for forested areas with heavy undergrowth.

Contours are not used on the military city map.

Red is used, on the civilian edition, only to provide the outline of federally owned property, and on one-way streets to indicate (by a red arrow) the traffic direction. On the military edition red is used to show installations of importance to the military.

In 1976 work on the NTS 1:25,000 series was reduced to the completion of those first editions already in work and the revision (on a six-year cycle) of sheets already published (see Figure 27). In 1978 work on the series was stopped altogether. The principal reason for this curtailment was the urgent demand for new 1:50,000 maps of the Canadian Arctic where the search for new oil and gas fields was being intensified. A secondary reason was to provide an opportunity to study the effects of the entry into mapping at similar scales by certain provincial mapping agencies. In 1979 British Columbia, Manitoba, Ontario and Quebec were mapping extensive areas with a monochrome 1:20,000 series. The Land Registration and Information Service in the Maritimes (see Chapter 13) is active at 1:10,000 and has made tentative plans for 1:20,000 mapping. Other provinces have indicated an interest in entering this field.

The military authorities, on the other hand, consider the Military City Maps series to be vital. The following is the current list of MCMs showing the date of the last edition:

MCE	Name	Civilian Edition Series A902	Edition	Validity date
335	Brantford		1	1975
315	Calgary		3	1975
336	Cambridge		1	1975
325	Charlottetown		1	1970
330	Chicoutimi		1	1974
344	Cornwall		1	1972
314	Edmonton		3	1976
323	Fredericton		2	1976

Table—*continued*

MCE	Name Civilian Edition Series A902	Edition	Validity date
337	Guelph	1	1975
301	Halifax	3	1975
309	Hamilton	3	1974
324	Kingston	1	1970
318	Kitchener/Waterloo	3	1972
311	London	3	1975
304	Montreal (Nord-North)	5	1973
305	Montreal (Sud-South)	5	1973
321	Montreal (Ouest-West)	2	1973
322	Montreal (Est-East)	2	1973
310	Niagara Falls	2	1970
331	North Bay	1	1974
343	Oakville	1	1974
332	Oshawa	1	1974
306	Ottawa/Hull	3	1975
328	Peterborough	1	1973
303	Quebec	4	1973
320	Regina	1	1972
342	Richmond Hill	1	1974
338	Sarnia	1	1975
339	Saskatoon	1	1975
333	Shawinigan	1	1974
334	Sherbrooke	1	1971
302	Saint John, NB	2	1972
300	St John's, NFLD	2	1971
319	Thunder Bay	2	1972
307	Toronto (East-Est)	3	1974
308	Toronto (West-Ouest)	3	1974
345	Trenton Belleville	1	1974
329	Trois-Rivieres	1	1973
316	Vancouver	4	1972
317	Victoria	2	1971
326	Whitehorse	1	1971
312	Windsor	3	1973
313	Winnipeg	4	1972
327	Yellowknife	1	1972

10. Maps of the Chief Geographer's Office

In 1890 the position of Geographer was created in the Department of the Interior. It was filled by the Chief Draughtsman of the Department and originally his duties were to produce small-scale maps of Canada as a whole, and regional maps that were required to illustrate the colonization and development projects of the Department. By the turn of the century a number of surveying and exploration projects were underway in Canada, all of them producing new information about the country in which they were operating and most of them also producing maps. An illustration of this and also of an early series produced by the Geographer is to be found in a 10-sheet set of maps on a scale of six miles to an inch covering the district between Wrangell, at the mouth of Stikine River, and the Porcupine River; i.e. the north-western portion of British Columbia and the western part of the Yukon Territory. They appeared in 1898 and were the most detailed maps of that part of Canada published up to that time. The maps were contoured insofar as the survey data would allow and the relief was enhanced with the use of hill shading. Sheet 1 also included a map of the town of Dawson on a scale of 600 feet to 1 inch. They were compiled from 'various reliable authorities': the Pacific coast and vicinity from the Canadian boundary surveys and the United States Coast and Geodetic surveys; the interior topography from surveys of the Department of the Interior and the Geological Survey and 'any other surveys and explorations that were available and authentic'.

But as a rule there was no coordination in the mapping, and consequently there was both duplication and gaps in the coverage. An attempt to produce an index of the maps of this period resulted in an almost unintelligible maze of lines illustrated in part in Figure 37 in Chapter 15.

The Chief Geographer's Series—1:250,000 and 1:500,000

In 1901 it was decided within the Department of the Interior that the Geographer's Office should commence production of two map series, one at 1:250,000, the other at 1:500,000,

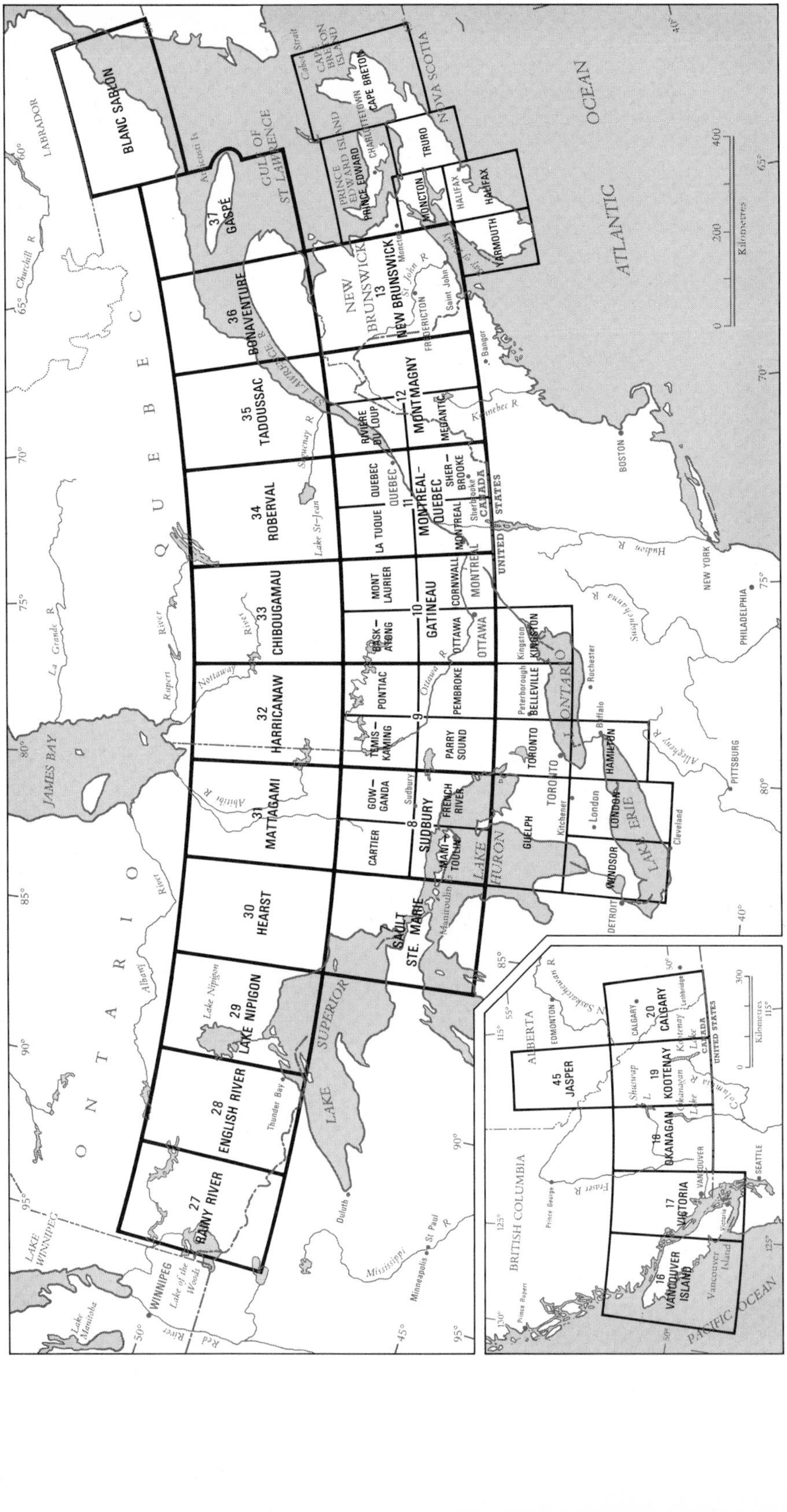

Figure 28 Index of Chief Geographer's series

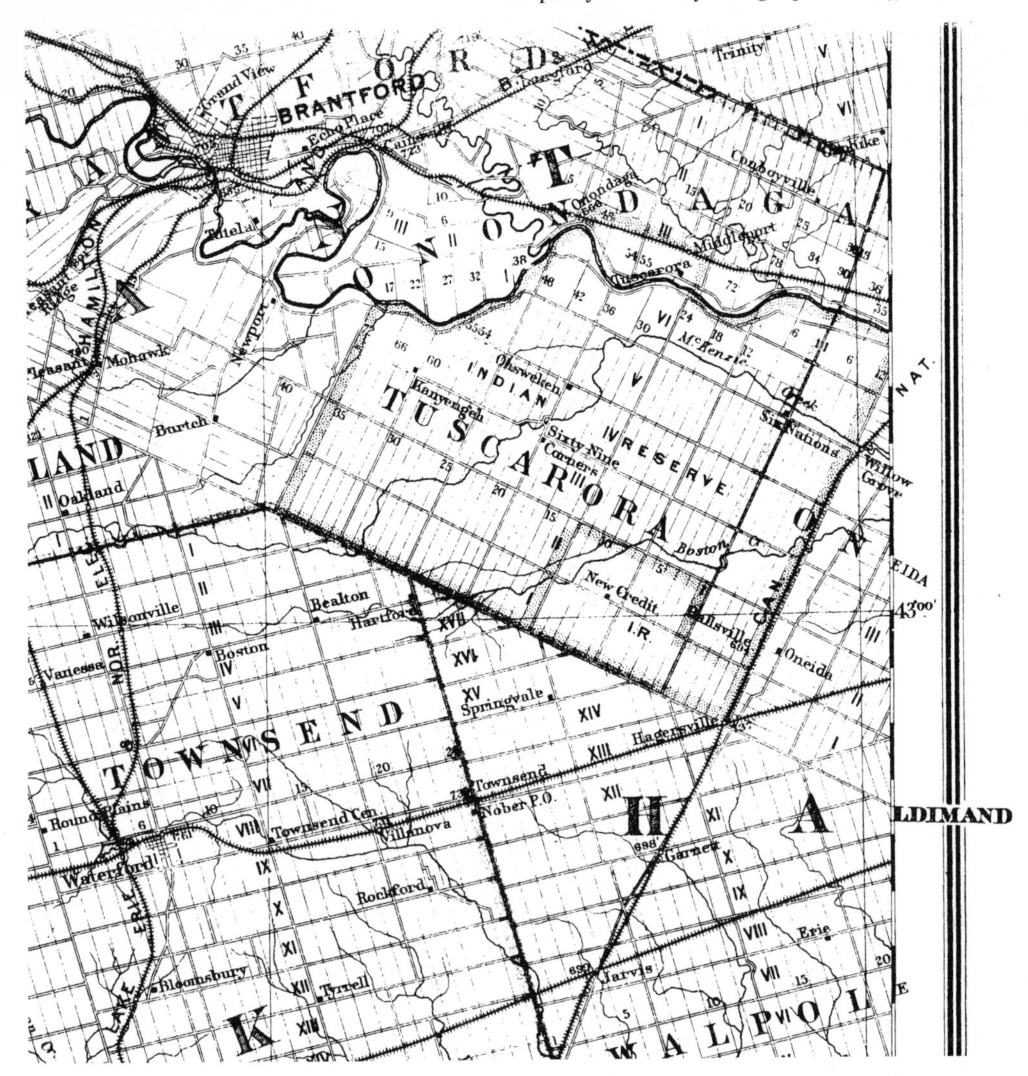

Figure 29 Section from a sheet of 1:250,000 Chief Geographer's series

which would distil all available cartographic information onto sheets with standard specifications and regular sheet lines. The basic sheets would be the 1:500,000 series which would be quartered into the 1:250,000 sheets. As shown in the index (Figure 28) the 1:500,000 sheets were 3° by 3°. Coverage was initially planned north to 54° (55° in western Canada) but presumably planners had in mind the eventual extension of the series across much, if not all of Canada. This expansion did not come to pass and in fact only 25 of the 1:500,000 series and 33 of the 1:250,000 series had been produced when it was realized that Canada had no need for such maps. The last new sheet, Pontiac (9 NE) was published in 1931. Up until 1931 revision was carried out conscientiously with new editions appearing every five or six years, but in the late 1930s sheets were reprinted without revision just to maintain the stock. Even this was discontinued in 1949 with the reprinting of Manitoulin (8 SW). Because of this reprinting policy many Canadian map libraries have a complete or almost complete set of these maps.[1] As they give a picture of a large portion of Canada as it existed in the inter-war years, a short review of the specifications of these series is justified. As both series are drawn in much the same style, and as a common set of symbols and abbreviations is used, the following remarks apply to both scales unless otherwise indicated. Figure 29 illustrates the style of both series.

Principal Features

COLOURS. Three colours are used in both series: black for all map detail, blue for open water, and orange for accentuation of international, interprovincial and county boundaries.

Table 21 Chief Geographer's Series 1:250,000

Number	Name	Editions
1 NW and NE	Guelph	1906, 1912, 1916, 1927, 1930, 1944
1 SE	London	1905, 1907, 1912, 1916, 1921, 1928, 1938
1 SW	Windsor	1904, 1907, 1912, 1915, 1916, 1923, 1928, 1936
2 NE	Belleville	1909, 1913, 1916, 1920, 1923, 1927, 1929, 1939
2 NW and 9 SW	Toronto, Muskoka	1907, 1912, 1914
2 NW	Toronto	1916, 1920, 1928, 1938
2 SW	Hamilton	1905, 1912, 1916, 1923, 1928, 1946
3 NW	Kingston	1911, 1915, 1919, 1924, 1928, 1937, 1946
4 NE	Yarmouth	1916, 1919, 1929, 1937
5 NE and 14 SE	Truro	1909, 1916, 1920, 1926, 1940
5 NW and 14 SW	Halifax	1910, 1912, 1916, 1920, 1927, 1929, 1937
5 NW and 14 SW	Halifax	1912 (showing) forest distribution
7 NW–NE	Sault Ste Marie	1912, 1918
8 NE	Gowganda	1910, 1919, 1921, 1925, 1929, 1944
8 NW	Cartier	1914, 1928
8 SE	French River	1916, 1920, 1927, 1946
8 SW	Manitoulin	1914, 1920, 1926, 1948
9 NE	Pontiac	1931
9 NW	Timiskaming	1908, 1919, 1924, 1928
9 SE	Pembroke	1909, 1915, 1919, 1921, 1923, 1927, 1931, 1942
9 SW	Parry Sound	1916, 1920, 1927, 1929, 1939
10 NE	Mont Laurier	1928
10 NW	Baskatong	1930
10 SE	Cornwall	1917, 1919, 1923, 1929, 1936
10 SE	Cornwall	1917 reprinted with electoral divisions
10 SW	Ottawa	1910, 1912, 1914, 1917, 1919, 1925, 1929, 1939
11 NE	Quebec	1922, 1929
11 NW	La Tuque	1928
11 SE	Sherbrooke	1923, 1929
11 SW	Montreal	1918, 1919, 1925, 1930
12 NW	Rivière-du-Loup	1929
12 SW	Megantic	1926
14 NE and 14 NW	Prince Edward Is	1940
14 SW (Part)	Moncton	1913
14 SW	Moncton	1918, 1925, 1937
14 SW and SE and 15 (parts of)	Prince Edward Is	1910, 1912, 1914, 1918, 1921 (2 sections), 1927, 1931, 1940
15 (parts)	Cape Breton	1907, 1916, 1920, 1925, 1929, 1937
31 SE	Abitibi	1910

Table 22 Chief Geographer's Series 1:500,000

Number	Name	Editions
7	Saulte Ste Marie	1927, 1938
8	Sudbury	1925, 1929, 1937
9	Nipissing	1922, 1927, 1937
10 and 33 (part)	Gatineau	1920
10	Gatineau	1924, 1927, 1929, 1939
11	Montreal and Quebec	1908, 1912, 1914, 1917, 1920, 1924, 1928, 1937
12	Montmagny	1913, 1918, 1923, 1937
12 and 35 (part of)	Montmagny	1913
13	New Brunswick	1905, 1912, 1919, 1923, 1928
16	Vancouver Island	1923, 1930
17	Victoria	1922, 1929
18	Okanagan	1923, 1936
19	Kootenay	1924, 1936
20	Calgary	1925
27	Rainy River	1905, 1914, 1918, 1923, 1927, 1930, 1941
28	English River	1909, 1914, 1918, 1928
29	Lake Nipigon	1907, 1917, 1922, 1927
30	Michipicotin	1910, 1918, 1924, 1928
	Hearst (see name change)	1928
31	Mattagami	1911, 1919, 1924, 1930, 1937
32	Harricanaw	1922, 1927, 1936
33	Chibougamau	1922, 1927, 1928, 1937
34	Roberval	1917, 1923, 1929
35	Tadoussac	1919, 1921, 1928, 1939
36	Bonaventure	1918, 1924, 1929
37	Gaspé	1918, 1924, 1929
38 (parts of 39, 64, 65)	Blanc Sablon	1921
45	Jasper	1926

CADASTRAL INFORMATION. The county and township structure is given great prominence in both series, both with the weight of boundary lines and the size and style of type of the county and township names. On the 1:250,000 sheets the full rural cadastre is shown, with each farm lot being outlined and every fifth lot numbered. This mass of detail overpowers all other depiction to as far north as the provincial land surveyors had completed this work at the time of the compilation of the map.

HYDROGRAPHY. Beyond the areas of the cadastral survey, the drainage becomes the prominent map feature. Rivers and lake shorelines are shown in black while the open water is in blue. Where river and lake information is uncertain (and there are many areas where this is so) the features are shown by pecked lines. In many cases lakes are left unconnected to any drainage system indicating that the unfortunate cartographer could find no data on the water flow in that area. On the Great Lakes and the oceans, where hydrographic charts were available, submarine contours are shown at 5-fathom intervals. One peculiar aspect of drainage depiction must be remarked upon. In surveying the wild areas to the north, it was the practice initially to run only the outlines of townships which would then be subdivided later when the settlers arrived. The surveyor would make a note of all streams crossed in the course of this perimeter survey, but as he only viewed the stream at the point he crossed it he had no idea of its source or destination. On the sheets of both series these sections of streams are shown as tiny arrows piercing the outline of the townships.

VEGETATION. Forest cover is not depicted in any way. On the 1:250,000 series wet ground is indicated by the standard marsh symbol.

RELIEF. No attempt has been made to show relief. Spot elevations are given to indicate the heights of mountains, and the elevations of railway stations and lakes are shown. But these series must be judged complete failures in their depiction (or lack of depiction) of the western mountains.

POPULATED PLACES. Cities are indicated by a conventionalized version of the street pattern. Towns and villages, regardless of size, are shown by a small open circle of 0.8 mm diameter. The population of cities and towns is indicated by the size of type used in the place name; the following classification being employed:

Cities
Towns over 4000
Towns under 4000
Villages

COMMUNICATION ROUTES. There is no doubt that the designers of these series considered the railway as the principal means of overland transportation. Rail lines are indicated by a black line with cross ties at 0.7 mm intervals, which makes them more prominent than other linear symbols except the county boundaries. Roads, on the other hand, are subdued. They are not classified, and therefore everything from a dirt road to a through highway is depicted by fine double black lines spaced 0.5 mm apart. This lack of classification makes both map series virtually useless as road maps. River routes have their own importance in wilderness areas, and consequently falls, rapids and portages are marked and identified by the appropriate abbreviation. Canals are shown by a double line similar to the road symbol but with an intermittent black fill. As canals are a rarity in the land covered by this series, the usefulness of this symbol is doubtful. A blue line with a label would have been much more to the point.

MARGINAL INFORMATION. The border is divided into black and white bars to represent one-minute intervals of latitude and longitude on the 1:250,000 series and five-minute intervals on the 1:500,000 series. Across the face of the map graticule lines are drawn at 15-minute intervals on the 1:250,000 series and 30-minute intervals on the 1:500,000 series. In the top margin the publishing department is given (The Department of the Interior before 1936, The Department of Mines and Resources after that date) and the date of publication and sheet number is stated. The side margins are bare except for latitude values and the exterior part of the names of counties that have been allowed to run off the sheet and which are completed in the margin in condensed type. (Obviously the designers expected sheets of the series to be joined to form larger wall maps.) A short legend of symbols and abbreviations appears in the bottom margin together with the sheet name, the provinces touched and the counties covered. A bar scale in miles and kilometres and the name of the Chief Cartographer completes the bottom margin.

The End of the Chief Geographer's Series

The annual report of the Surveyor-General for 1901 includes Sessional Paper No. 25 by Chief Geographer James White. This paper describes the planned 1:250,000 series and supports the view that this was a determined attempt to make order out of the chaos that existed in the mapping of Canada at the beginning of the twentieth century. Although White does not make the point, there must have been an almost primitive urge to start getting Canada mapped at scales larger than those used by explorers, to start putting on paper both what was known in the settled areas and what was suspected or even rumoured in regions more remote. But these series did not survive the competition for map-users provided by the series of the National Topographic System.

The NTS four-mile series was a true topographic map which also was useful as a road map and the NTS eight-mile series was proving useful as a base for Canada's first aeronautical chart series. The Chief Geographer's maps, with their emphasis on cadastral surveys and their complete disregard for emerging forms of transportation, had no strong group of users ready to demand their continued existence. Consequently they were abandoned.

Although they are gone these maps are not completely forgotten, and their influence lives on in subtle ways. Three provinces, Ontario, Quebec and British Columbia, publish small-scale cadastral maps that perpetuate some of the symbolization of the 1:250,000 series (Quebec publishes at 1:200,000, British Columbia and Ontario at 1:250,000). The cartographic method used by the Chief Geographer to accentuate county boundaries, (a line of solid colour immediately inside the boundary together with an additional broader line of a screened tint of the same colour) has been used on the present 1:2,000,000 map of Canada.

But the Chief Geographer's Office was also involved in the beginning of another important map series. Ever since 1877 the Office had prepared special maps for international conferences in which Canada was involved but the International Map of the World was by far the most important of these activities.

The IMW 1:1,000,000 Series

The formal proposal to produce an International Map of the World (IMW) at the scale of 1:1,000,000 was made by Professor Albrecht Penck, a German geographer, at the Fifth International Geographical Congress in Switzerland in 1891. At the Eighth International Geographical Congress held at Washington in 1904 he appealed to the United States and Canada 'to undertake this work in so far as it affected their respective dominions' and the Geographer of the Department of the Interior who attended this Congress subsequently strongly recommended it. He pointed out that maps on this scale would be useful to the Railway Commission of Canada and that they could be utilized to form maps of the various provinces which would include the whole of each province and at the same time would not be so large as to be unwieldy. Some work was started in 1905 and the Chief Geographer (as the Geographer was now called) attended the meeting of the First International Map Conference in London in 1909 when general specifications and production methods were agreed upon. The Second International Map Conference held in Paris in 1913 approved detailed specifications, but the First World War and its aftermath delayed rapid progress.

Table 23 IMW Series 1:1,000,000

IMW no.	Name	Edition	
NK 17	Lake Erie	49–01	74–02 US
NK 18	Hudson River		61–01 US
NL 16	Lake Superior		66–01 US
NL 17	Sudbury		70–01
NL 18	Ottawa	29–01, 64–02,	67–03, 74–04
NL 19	Quebec		68–01 US
NL 20	Halifax	31–01	69–02
NL 21/22	St Johns		70–01
NM 9/10	Vancouver		78–01
NM 11	Kootenay Lake	69–01	78–02
NM 12	Lethbridge		73–01
NM 13	Regina	28–01	70–02
NM 14	Winnipeg		70–01
NM 15	Lake of the Woods		70–01
NM 16	Lake Nipigon		72–01
NM 17	Timmins		71–01
NM 18	Lac Chibougamau		71–01
NM 19	Chicoutimi		71–01
NM 20	Anticosti Island		69–01
NM 21/22	Corner Brook		69–01
NN 8/9	Prince Rupert		70–01
NN 10	Prince George		70–01
NN 11	Lesser Slave Lake		71–01
NN 12	Edmonton		73–01
NN 13	Prince Albert		71–01
NN 14	The Pas		72–01
NN 15	Gods Lake		70–01
NN 16	Winisk Lake		70–01
NN 17	James Bay		70–01
NN 18	Lac Bienville		71–01
NN 19	Schefferville		71–01
NN 20	Churchill Falls		73–01
NN 21	Cartwright		70–01
NO 7/8	Sitka		76–01 US
NO 9	Dease Lake		71–01
NO 10	Fort Nelson		77–01
NO 11	Peace River		77–01
NO 12	Lake Athabaska		75–01
NO 13	Wollaston Lake		75–01
NO 14	Southern Indian Lake		75–01
NO 15	Churchill		77–01
NO 16	Partridge Island		76–01
NO 17	Ottawa Island		76–01
NO 18	Lac Minto		77–01
NO 19	Ungava Bay		79–01
NO 20	Torngat Mountains		73–01
NP 7/9	MacMillan River		75–01
NP 9/10	Redstone River		75–01
NP 11/12	Slave River		75–01
NP 12/13	Lockhart River		75–01
NP 13/14	Dubawnt Lake		75–01
NP 15/16	Maguse River		75–01
NP 16/17	Sutton River		78–01
NP 17/18	Riviere Kovic		79–01
NP 19/20	Soper River		79–01
NQ 7/9	Peel River		76–01
NQ 9/12	Great Bear River		76–01
NQ 12/14	Thelon River		76–01

IMW no.	Name	Edition
NQ 15/17	Quoich River	79–01
NQ 17/20	Koukdjuak River	79–01
NQ 20/22	Davis Strait	80–01
NR 7/9	Firth River	76–01
NR 9/12	Horton River	76–01
NR 12/14	Victoria Strait	76–01
NR 15/17	Murchison River	80–01
NR 17/20	Rowley River	80–01
NS 9/12	Thompsen River	76–01
NS 12/14	Viscount Melville Sound	76–01
NS 15/17	Lancaster Sound	76–01
NS 17/20	Eclipse Sound	78–01
NT 8/12	Ballantyne Strait	76–01
NT 12/16	Belcher Channel	76–01
NT 16/20	Jones Sound	77–01
NU 14/20	Robson Channel	78–01

However these developments were not lost on Canada for when the Board of Topographical Surveys and Maps decided that it was essential that the mapping of Canada should be coordinated into an orderly system suitable for series maps of different scales and for coverage of the whole country, the first indexing of the National Topographic Series was based on the numbering of primary quadrangles corresponding to the 'Carte du Monde'. Under this plan, adopted in 1925, each primary quadrangle covered 6° of longitude and 4° of latitude, the recommended dimensions for the IMW.

This sheet arrangement was abandoned by Canada for its national mapping system in 1927 but three 'experimental' sheets were produced in accordance with IMW specifications—NW 13 Regina was drawn by the Topographical Survey and published in 1928[2]—in time to be exhibited at the Twelfth International Geographical Congress in Cambridge, England. NL 18 Montreal and NL 20 Nova Scotia were produced in 1929 and 1931 respectively by the Chief Geographer's Office. The sheet lines of NL 20 were modified slightly in the southwest and northeast so as to accommodate the whole of peninsula Nova Scotia and Cape Breton Island.

Further progress on the IMW series was also delayed by the need to produce aeronautical charts on the scale of 1:1,000,000, especially during and after the Second World War, and the concern that the International Map of the World would duplicate the functions of the World Aeronautical Chart. In 1962, as the result of an international conference held under the auspices of the United Nations in Bonn, it was decided to keep the two series separate because their objectives are different. The detail on the WAC sheets is concerned with assisting air navigation while that on the IMW sheets is for research purposes and assisting economic development. However, modifications to the IMW specifications were also agreed upon which would bring the two series closer together. For example the projection originally recommended for the IMW was a modified polyconic which is neither conformal or equivalent. The WAC, on the other hand, used the Lambert conformal conic which has small-scale error, relatively straight azimuths over the extent of a single chart and correct shapes around a point. But the differences between the two projections are not noticeable on casual inspection and the Bonn conference agreed that this projection could be used for IMW sheets. Standard parallels are specified for each four-degree band (being

40 minutes south of the north border, and 40 minutes north of the south border of each sheet). It also confirmed the original intention to depart from the six-degree longitude width north of 60° north. Since the convergence of the meridians made such sheets very narrow, two or more adjacent sheets could be published as one.

These changes suited Canada very well and in 1964 a second edition of NL 18 was published with the title changed from Montreal to Ottawa. It was revised and issued as a third edition in 1967. Clearly, by this time the decision had been made to proceed with the series as in that year the old Nova Scotia sheet NL 20 was revised and renamed Halifax; the Regina sheet NM 13 was also revised and seven new sheets were compiled. By 1970, eight sheets had been published.

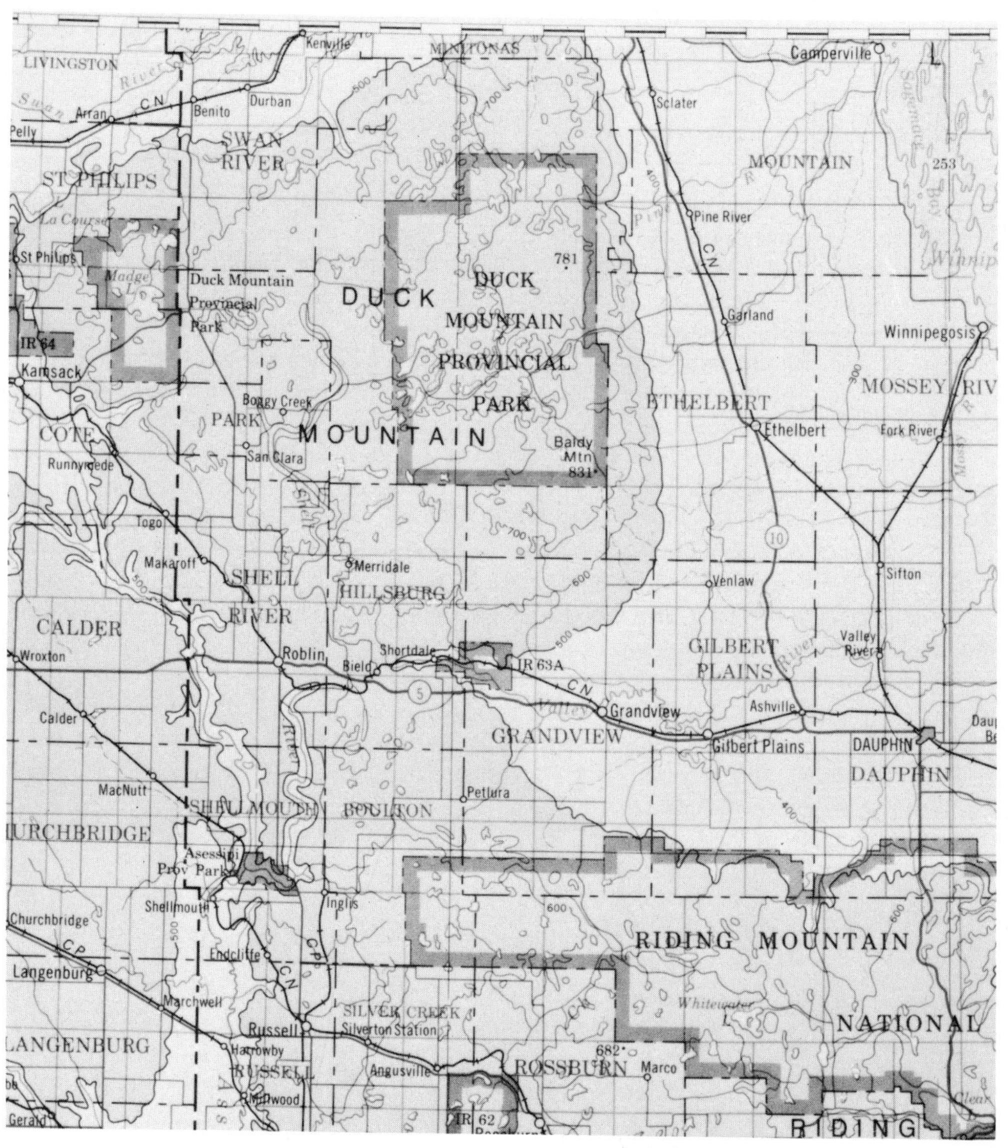

Figure 30 Section from a 1:1,000,000 IMW sheet

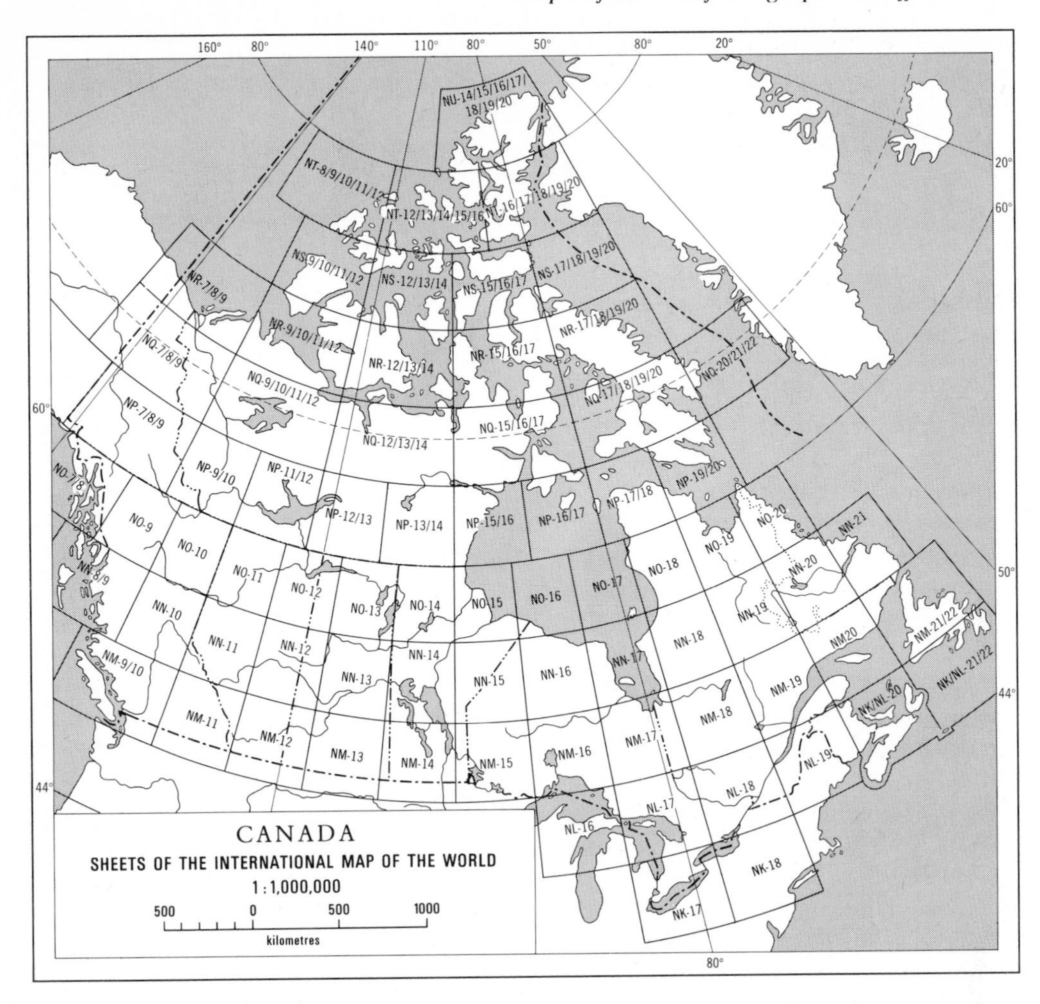

Figure 31 Index of IMW coverage of Canada

New Sheets Published by Canada in IMW Series

Before 1968	3
1968–69	5
1970–71	16
1972–73	7
1974–75	11
1976–77	15
1978–79	6
1980	6
	69

There are five additional sheets which cover parts of Canada which have been published by the USA in accordance with the agreement under which the country which occupies most of any sheet is responsible for its preparation. Thus by 1980 all of the IMW sheets had appeared. The speed with which this series was completed was undoubtedly because the 1:50,000 and the 1:250,000 mapping was well advanced by 1970 and the IMW series was compiled from the maps at these scales. Figure 30 is a section from sheet NM 14 Winnipeg, while Figure 31 shows the layout of IMW sheets covering Canada.

The hypsometric system of relief is used with principal contours at 100, 200, 500, 1000, 1500, 2000, 2500, 3000 metres (then up by 1000's of metres) above sea level, the tints between progressing from green in the lowlands through yellow, brown and purple for the mountains. Metre contouring is obtained from foot-contoured source material by interpolation. Principal isobaths for the water areas are drawn at 100, 200, 500 and 1000 metres and then at 1000-metre intervals below sea level, the tints of blue between them becoming progressively deeper the deeper the water. For depths below 6000 metres purple is used.

In addition, particular relief, hydrographic and snow and ice features, are shown by conventional symbols as are populated places, roads, railways, the boundaries of provinces, districts, counties, municipalities and townships, Indian reserves, national and provincial parks, forest reserves, and prominent industrial and public works features such as pipelines and airports.[3]

The sheet reference system was designed for use without an index. It consists of two letters followed by a number. The first letter indicates the hemisphere, hence all Canadian sheets begin with the letter N. The second letter indicates the four-degree band north of the Equator. The southernmost of these bands which cover Canada is the 11th (40° to 44°) and hence is designated by the letter K. The northernmost of the band (80° to 84°) is designated by the letter U. The map sectors, 6° in width, are numbered eastward around the earth from the 180th meridian. These numbers begin with 7 in western Canada and extend to 22 in the east. They follow the appropriate letters.

Within the map a further reference grid is provided by the use in the vertical margins of lower case letters a to h, reading from top to bottom for each 30 minutes of latitude, and in the horizontal margin Roman numerals, reading from left to right for each 30 minutes of longitude. The IMW system offers an efficient breakdown into larger scales that have been adopted by the Canadian JOG series (see Chapter 8) and by many countries for their whole topographic system. In retrospect, it is somewhat sad that Canada tried the IMW numbering system and rejected it in 1927. The attraction of the 8° Transverse Mercator was probably to blame.

11. Federal Thematic Maps[1]

When the IMW was initiated it was envisaged as a general planning map which would serve as a base for maps of other distributions—population, ethnic groups, archaeology, vegetation, soils and geology—in other words for thematic maps. But even if Canada had produced more than three IMW sheets by 1966 it is doubtful if they would have been used in this way partly because the need for and hence the production of some thematic map series long antedated the IMW project and partly because of the constraints of the scale of 1:1,000,000.

Geology

As has already been pointed out geological maps were needed in regions where prospecting and mining activities were most pronounced and this frequently meant that the geologists had to produce their own topographical base on which to superimpose thematic information. The differences between the base maps produced by the topographical division of the Geological Survey of Canada and the other federal mapping agencies have been described in Chapter 4.

The cartographic preparation of maps produced by the Geological Survey of Canada at scales of 1:50,000, 1:250,000, 1:500,000 and 1:1,000,000 is now standardized. For larger and smaller-scale thematic maps the standards are applied in a more general manner. The geological information which overlays the bases has become codified over the years. A list of more than 80 standardized geological symbols has been developed over a period of 25 years. In addition the maps use particular colours and colour tones along with letters to show the extent of different classes of rock.

Geological maps and indexes are available from the Publications and Information Office, Geological Survey of Canada.

Soils

Soils maps on topographic scales are produced by the provincial soil scientists in cooperation with federal agencies. In Ontario, Quebec, Nova Scotia and Prince Edward Island they cover counties. In the remaining provinces and the territories they cover special areas or regions or map sheets. The first soil map in this series was published in 1928. It was for Norfolk county in Ontario. The soil data was the result of surveys carried out by the soil scientists at what is now the University of Guelph in 1927–28. It was mapped on base maps of the then federal Department of Militia and Defence and published by the federal Department of Agriculture on a scale of two miles to one inch. Most of the maps up to 1945 were published on the same scale: thereafter they are on the one mile to one inch scale, exceptions being Bruce County which was on the scale of 1½ miles to one inch and Parry Sound on two miles to one inch. The last soil map to be published was for Halton County in 1971.

The scales are typical of those used for all of the eastern provinces and the earliest sheets of British Columbia. Later sheets for British Columbia have been on a scale of 1:126,720 (two miles to one inch). In Manitoba and Saskatchewan the scales are almost all two miles to one inch while in Alberta although the two-mile scale is favoured some sheets are on the three-mile and 12-mile scale. Similarly the published maps for the Yukon and Northwest Territories have varied in scale from 1:125,000 to 1:506,880.

The contours of the topographical base are printed in brown and are easily readable through the colour patches used for the various soil types. Each soil type is assigned a colour and one or two letters, and both colour and letters appear on the map. In some of the more recent maps the soil symbol letters are followed by a third letter indicating a topographical slope according to a 14-point scale and a figure indicating the degree of stoniness (on a 5-point scale).

Soils maps are available from the provincial Departments of Agriculture in Prince Edward Island, Quebec, Manitoba, Saskatchewan, Alberta and British Columbia; in Newfoundland from the Department of Mines, Agriculture and Resources; in New Brunswick from the Department of Agriculture and Rural Development; in Nova Scotia from the Department of Agriculture and Marketing and in Ontario from the Ministry of Agriculture and Food. Soils maps of the Territories are obtainable from Agriculture Canada in Ottawa.

Vegetation

Another of Canada's great economic resources is its forests. Vegetation has always been mapped on the topographical maps produced in Ottawa (except those produced by the Geological Survey before 1936) but the use of topographical maps as bases on which to superimpose details of the forest cover dates from 1921 and depended very much on the development of the techniques of aerial photography and the interpretation of forest detail from the resulting photographs. As with geology, the greatest problem in the early years was the lack of satisfactory base maps of the area being flown. But forest type maps covering about 15,000 square miles were completed for northern Ontario. As with soils maps, the production of the forest cover maps was frequently a cooperative venture between the federal government and the government of the province concerned. One of the earliest maps to be produced in colour appeared in 1941 on a scale of eight miles to one inch.

Shortly after the Second World War a forest inventory series was begun on a scale of one mile to one inch. The maps showed the type of forest (softwood, hardwood or mixed) by means of an appropriate colour patch, the height being differentiated by the intensity of the colour. Overprinted in black were letters denoting the percentage of crown cover, figures denoting crown cover height and symbols indicating special forest features such as 'recent burn' or 'recent windfall'. A number of map sheets covering scattered areas were completed before the programme was taken over by each of the provinces. However in the mid-1960s a similar series was begun for the Northwest Territories and the Yukon on a scale of 1:250,000. It was called the Forest Cover series and although several map sheets reached the pre-printing stage, only one map, Coal River (NTS 95D), was actually published. The series was discontinued after a survey among all potentially interested parties indicated that there was not sufficient demand to justify the effort.

Table 24 Land Use Series

Name	NTS number	Scale	Dates
Halifax	11 D	1:250,000	1961
Truro	11 E	1:250,000	1961
Canso	11 F	1:250,000	1961
Sydney	11 K	1:250,000	1962
Shelburne	20 O	1:250,000	1961
Annapolis	21 A	1:250,000	1964
Eastport	21 B	1:250,000	1966
Fredericton	21 G	1:250,000	1964
Amherst	21 H	1:250,000	1964
Kings	21 L/SE	1:126,720	1963
Prince	21 I/NE	1:126,720	1962
Queens	21 L/SW	1:126,720	1963
Dunnville E	30 L/13	1:50,000	1961
Dunnville W	30 L/13	1:50,000	1963
Welland E	30 L/14	1:50,000	1962
Welland W	30 L/14	1:50,000	1961
Fort Erie W	30 L/15	1:50,000	1963
Niagara E	30 M/3	1:50,000	1964
Niagara W	30 M/3	1:50,000	1961
Grimsby E	30 M/4	1:50,000	1962
Grimsby W	30 M/4	1:50,000	1963
Pelee	40 G/15	1:50,000	1962
Long Point W	40 I/9	1:50,000	1964
Simcoe E	40 I/16	1:50,000	1963
Simcoe W	40 I/16	1:50,000	1966
Essex E	40 J/2	1:50,000	1963
Brantford E	40 P/1	1:50,000	1962
Moose Jaw–Watrous	72 NE	1:500,000	1962
Hanna–Kindersley	72 NW	1:500,000	1964
Sooke	92 B/5	1:50,000	1963
Victoria	92 B/6	1:50,000	1962
Sidney	92 B/11	1:50,000	1963
Shawnigan	92 B/12	1:50,000	1962
New Westminster	92 G/2	1:50,000	1963
Vancouver South	92 G/3	1:50,000	1963
Chilliwack	92 H/4	1:50,000	1964

Land Use

Maps which brought together all of the actual uses of the land at a given time began to appear after the Second World War, perhaps partly inspired by the Land Utilization Survey of Britain. A land-use map of the Peace River District of British Columbia was published by the provincial government in 1948 at a scale of two miles to one inch and a similar map of the Terrace District of British Columbia on the one mile to one inch scale in 1950. From 1961 to 1966 an extensive series of such maps was published by the federal government. The land-use categories conform to the recommendations of the International Geographical Union. There are sixteen major classes each with its own distinctive colour or colour shade. For example, reds are used for urban (non-agricultural) uses; browns for grain growing; purples for orchards and horticulture; greens for improved pasture and orange for open grassland and scrub grassland. The topographical base information is printed in grey but is readable through the land-use colours.

The scales vary, (see Table 24). The seven sheets which together cover the whole of Nova Scotia are on a scale of 1:250,000, as are the only two New Brunswick sheets. The three sheets covering Prince Edward Island are on a scale of 1;126,720. The fifteen sheets covering parts of southern Ontario and the seven covering parts of British Columbia are on the 1:50,000 scale. Two sheets were produced for the Prairie region, both on a scale of 1:500,000. Present land-use data is no longer published in map form but is made available as computer mapping inputs. This series is therefore only available in the map libraries of government departments and universities.

Land Capability

A comprehensive survey of assessments of land capability for agriculture, forestry, recreation and wildlife was undertaken from 1963. It includes a series of maps on each of these themes (two in the case of wildlife—one for waterfowl and one for ungulates), which cover an area of approximately one million square miles including all of the Maritime Provinces, the island of Newfoundland, and the settled portions of Ontario, Quebec and the Western Provinces. The provinces carry out the work within their own boundaries. Collectively the maps are referred to as the Canada Land Inventory. They are published on three scales, 1:50,000, 1:250,000 and 1:1,000,000, using the NTS maps as the bases. The 1:50,000 maps are used as basic documents for planning and are available only from the provinces they cover. The data are generalized for the 1:250,000 and 1:1,000,000 series and published by the federal government in colour. Those available at 1:1,000,000 are shown in Table 26. It should be noted that on this scale maps are available which show critical capability areas. These are areas which represent the lands that are the most productive, or the best suited to a range of uses, or that are essential for the maintenance and production of wildlife.

Agriculture

The mineral soils are grouped into seven classes and 13 subclasses according to the potential of each soil for the production of field crops. Organic soils are shown as a single separate unit. The classes which are designated by different colours indicate the degree of limitation imposed by the soil in its use for mechanized agriculture. Class 1 soils have no limitations. Soils of Classes 2 and 3 have moderate and moderately severe limitations.

Class 4 soils are marginal for sustained arable agriculture. Class 5 soils are best suited only for the continuous growth of hay or pasture and Class 6 soils only for grazing. Class 7 soils are considered unsuitable for agriculture.

The following subclasses, designated by a letter, indicate the kinds of limitations that individually or in combination are affecting agricultural land use: climate (C), poor soil structure (O), erosion damage (E), low fertility (F), inundation (I), low water holding capacity (M), salinity (N), stoniness (P), shallowness to bedrock (R), numerous undesirable soil characteristics (S), adverse topography (T), excess water (W) and the cumulative effect of numerous minor limitations (X).

Forestry

Land is rated according to seven classes depending upon its capability to grow commercial timber in areas stocked with the optimum number and species of trees without improvements such as fertilization and drainage. The capability class is indicated by large Arabic numerals and a distinctive colour, Class 1 (dark green) being the best lands for tree growth and Class 7 (purple) the lands which cannot yield timber in commercial quantities. Smaller numerals following the capability class give the proportion of that class out of a total of ten. The capability subclass, shown as a capital letter, indicates the environmental limitations to tree growth. Letters under the capability class represent the native tree species that can be expected to yield the volume of timber associated with the capability class.

Recreation

Recreational land use capability is also indicated by Arabic numerals ranging from 1 to 7, each category having its distinctive colour. Subclasses are indicated by letters following the number (see Table 25).

Table 25 Recreation Subclasses

A	Angling
B	Beach
C	Canoeing
D	Deep inshore water
E	Recreationally significant vegetation
F	Waterfalls or rapids
G	Glacier view or experience
H	Historical or archaeological interest
J	Gathering and collecting
K	Organized camping
L	Landforms
M	Small surface waters
N	Lodging (cottaging)
O	Upland wildlife
P	Cultural landscape patterns
Q	Variety in topography or activities
R	Rock formations
S	Skiing
T	Thermal Springs
U	Deep water boating
V	Viewing
W	Wetland wildlife
X	Other features
Y	Family boating
Z	Man made structures

Wildlife

Both wildlife series use the same mapping methods, i.e. a seven-class capability classification which indicates the degree of limitation for the survival, growth and reproduction of wildlife. Each class has its distinctive colour patch. A subclassification uses a letter to indicate the factors that cause the limitation.

In the case of the ungulates, letters beneath the class numeral indicate the species to which the capability class refers, e.g. M for moose, E for elk and D for deer.

Land Use Planning

A final series is derived from the preceding series by mapping the areas of Classes 1, 2 and 3 for agriculture, forestry, recreation and ungulates so as to determine the distribution and acreage of the presumed best areas for each land resource activity. After the high potential areas are plotted the Class 4 lands for agriculture and forestry are added followed by the areas suitable for perennial forage. Then the large urban areas are indicated and the remaining land is classified as having low potential for all land resource uses. Once again each derived area has its distinctive colour patch and identifying letter.

Table 26 The 1:1,000,000 Land Capability Maps

Map title	Forestry	Agriculture	Wildlife (ungulates)	Wildlife (waterfowl)	Recreation	Critical capability areas
Atlantic Provinces (New Brunswick Prince Edward Island, Nova Scotia and the Island of Newfoundland)	X	X		*	X	X
Quebec (south)		X	X	X	X	
Ontario (south)	X	X	X	X	X	X
Manitoba (south)		X	X	X	X	X
Saskatchewan (south)	X	X	X	X	X	X
Alberta	X	X	X	X	X	X
British Columbia (north)				X	X	
British Columbia (south)				X	X	

* Except the Island of Newfoundland.

Northern Land Use Information

The Land Capability series just described covers only the southern part of Canada. For the north the governments of the Northwest Territories and the Yukon began in 1972 a programme of publishing land use information on maps on a scale of 1:250,000. They summarize information on renewable resources and related human activities in order to facilitate comprehensive regional planning and a managed approach to development and environmental protection. The legend and explanatory notes and diagrams occupy more than half the area of each sheet. A legend summary is given in Figure 32. Up to 1978, 162 sheets had been published covering all of the Yukon and all of the Mackenzie valley. They are available from the Canada Map Office.

LEGEND SUMMARY

This legend summarizes the symbols used in the map series with the exception of standard topographic symbols, which are printed along the bottom margin of each map sheet. In general, wildlife information is shown in red; hunting, trapping and fishing information in blue; recreation and other information in black. Detailed explanations of symbols will be found in subsequent sections of this map.

SYMBOLS

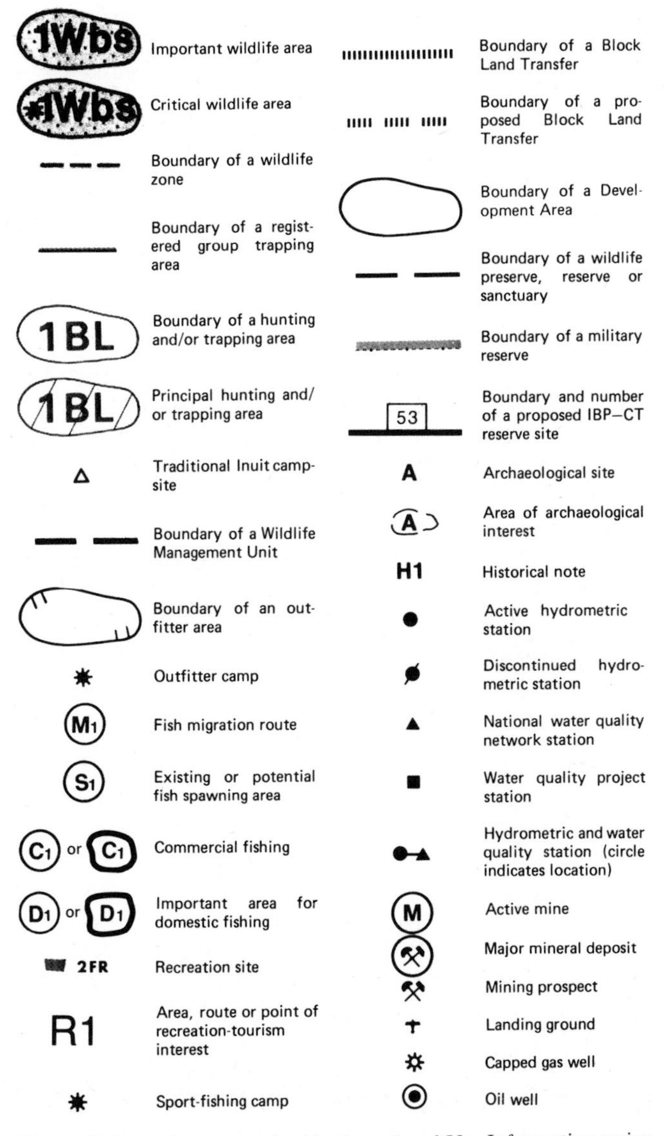

Figure 32 Legend summary for Northern Land Use Information series

Arctic Ecology

Related to the Northern Land Use Information series is the Arctic Ecology Map series the purpose of which is to record important and critical wildlife areas of the Canadian Arctic where human activities can have an adverse or destructive impact on wildlife populations. It brings together as much data as possible on habitats utilized by a wide range of species.

The base maps for four of the sheets are black and white prints of the NTS 1:500,000 series. However, for most of the sheets (98 in 1978) the bases are black and white prints of the ICAO 1:1,000,000 World Aeronautical Chart. The wildlife information is overprinted in black. The boundaries of the critical and important areas are shown and within them a figure-letter code gives particular information, e.g. 1Wbs, where 1 is the unit description number, W the species (in this case waterfowl), b the habitat function (in this case breeding) and s the seasonal utilization (in this case summer only).

Natural Resource Maps

Another series superficially similar to the Land Capability series has been started recently. The maps are referred to as natural resource maps, but the title is misleading as they actually show the distribution of some of the physical characteristics of the land beneath Canada's territorial seas. All the maps are on a scale of 1:250,000 and most cover a 1° by 2° area indexed and referenced according to a system devised by Marsden in 1831 mainly for identifying the geographical position of meteorological data over the oceans.

The whole surface of the earth is divided into 'Marsden Squares', 10° of latitude by 10° of longitude. Each 'square' is given a three-digit identifying number. The square is again divided into 100 one-degree squares (10 by 10) which are numbered starting with 00 in the southeast corner and progressing to 99 in the northwest corner. In this way every one-degree square on the earth's surface is described uniquely by a five-digit number. South of 60° the natural resource maps cover the area of two one-degree squares, and each map takes its number from the right-hand one-degree square. North of 60° the maps cover three, and in the far north four, one-degree squares, but they always take their number from the right-hand square, as shown in Figure 33.

There are or will be six thematic maps for each map area. These are identified by a letter following the Marsden number according to the following code:

A. Bathymetry
B. Gravity (free air anomaly)
C. Magnetism (total field)
D. Gravity (Bouguer anomaly)
E. Magnetism (anomaly)
F. Fisheries chart

In addition a plotting base, which is a print of the published bathymetry edition, is available in grey, and can be used as a working chart.

Most of the sheets published so far cover the Gulf of St Lawrence and the seabed off the Atlantic Provinces, but there are some covering the Pacific coastal area south of Queen Charlotte islands, five off the Mackenzie Delta and six spanning the eastern section of the Parry Channel. They may be obtained along with an index from the Hydrographic Chart Distribution Office, Environment Canada.

National Parks

The National Parks series is thematic only in that it consists of maps of national (federal) parks. Otherwise it could be considered as a series of special topographical maps, 'special' because the sheet lines do not fit the NTS index, but are determined solely by the dimensions of the park area.

The preparation of such a special series began with the appearance of the Kootenay and Waterton Lakes sheets in 1928. It was undoubtedly stimulated by the passage of the

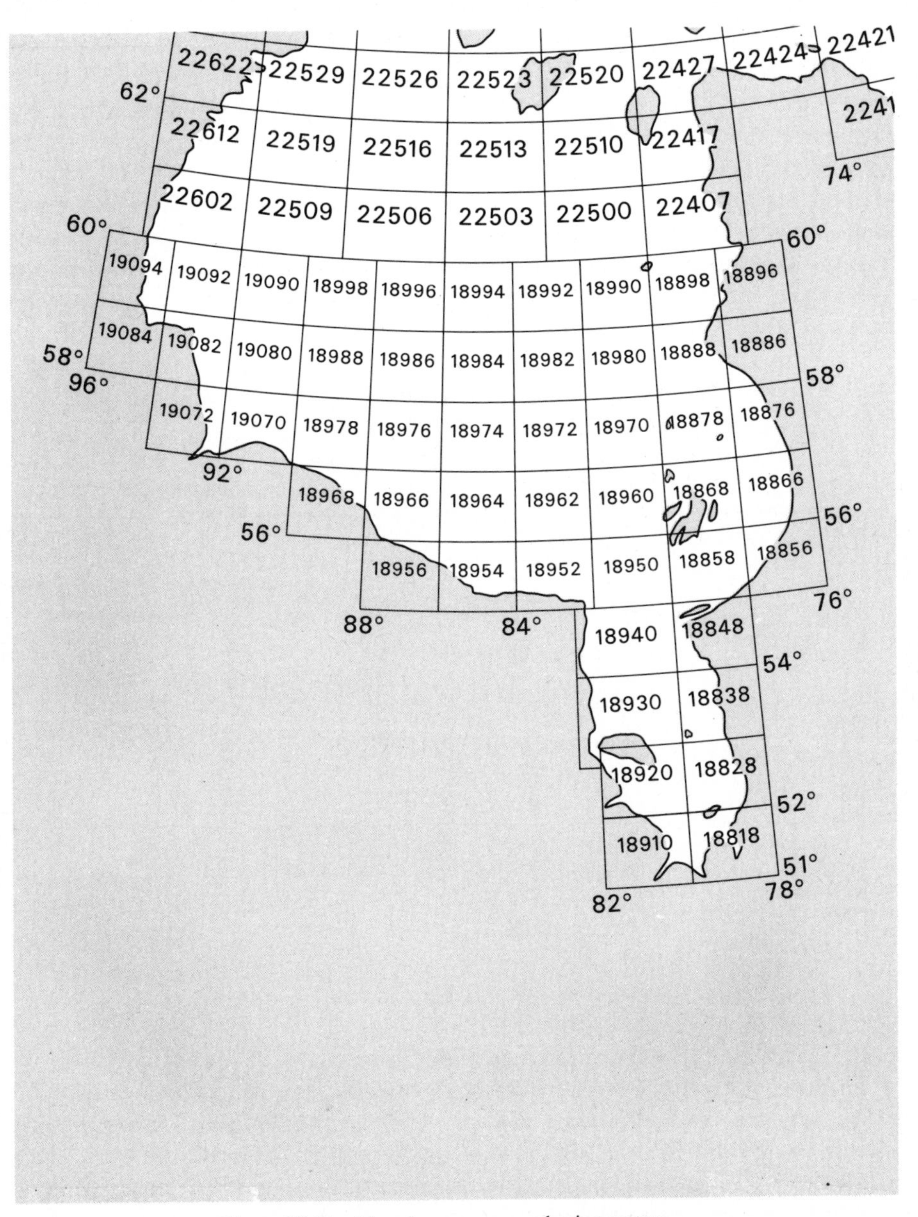

Figure 33 The Marsden square numbering system

National Parks Act of 1930. The Yoho sheet was published in that year; Prince Albert in 1931; Banff and Riding Mountain in 1932 and Glacier and Jasper in 1934 after which the publication of new sheets came to an abrupt halt (although new editions of already published sheets continued to appear). With the exception of Waterton Lakes at one mile to one inch (1:63,360) the scales were between two miles to one inch (1:126,720) and three miles to one inch (1:190,080). There was little consistency in the cartographic specifications for these maps apart from the depiction of man-made features. In the western Cordillera the maps were contoured, but some of them (such as Banff and Jasper) showed no vegetation, neither by symbols nor by colour patches. The maps of the parks in the Great Plains (Prince Albert and Riding Mountain), however, were not contoured, but did show wooded areas either by a uniform green colour or by green tree symbols. One new map in the series was published shortly afterwards in 1947 when the Wood Buffalo sheet appeared. It was, however, only a preliminary edition, printed in black and water blue only, without contours and on a scale of eight miles to one inch (1:506,880).

The years following saw the establishment of a number of additional parks and with it a review of the map series. The first editions of the maps of 'new' parks (such as Fundy and Terra Nova) and the latest editions of the maps of many of the 'old' parks are contoured with coloured shaded relief; they usually include large-scale insets of the park headquarters area and all are on metric scales.

Table 27 National Parks Series

Name of park	Date of first edition	Scale of first edition	Date of latest edition	Scale of latest edition
Waterton Lakes	1929	1:63,360	1973*	1:50,000
Kootenay	1928	1:126,720	1974	1:126,720
Yoho	1930	1:126,720	1961	1:126,720
Prince Albert	1931	1:150,000	1973*	1:125,000
Banff	1932	1:190,080	1974	1:190,080
Riding Mountain	1932	1:190,080	1976*	1:125,000
Glacier	1934	1:126,720	1974	1:126,720
Jasper	1934	1:190,080	1974	1:250,000
Wood Buffalo	1947	1:506,880	1947**	1:506,880
Fundy	1962	1:50,000	1962*	1:50,000
Mount Revelstoke	1963	1:50,000	1971*	1:50,000
Terra Nova	1966	1:50,000	1966*	1:50,000

* Shaded relief editions.
** Out-of-print. For this park and some 15 national parks which are not covered by special maps, standard topographical sheets must be used.

Road Maps[2]

Canadian official road maps are produced annually or biennially by each of the provinces and territories, and by the federal government. In each case the producing agency is that concerned with tourism. Although they vary widely in scale (which is understandable considering the great differences in areas of the regions concerned), there is a similarity which enables a person familiar with the road map of one province to become almost instantly 'at home' with that of another. To begin with, the road hierarchy is similar in all

Table 28

Province, territory or Canada		Approximate scale of main map	Size of map in cm, east–west dimension given first	Language of map	Unit of distance	Relief depiction	Remarks
Alberta		1:500,000	46.0 × 83.8	English only	Kilometres	None	
British Columbia		1:2,400,000	67.3 × 66.5	English only	Kilometres and miles	Only mountain peak heights (in metres)	The lack of shaded relief gives this very mountainous province a strangely flat appearance on this map
Manitoba		1:900,000	54.6 × 101.6	English only	Kilometres	Some hill shading	Manitoba is very flat, so the hill shading is a brave attempt at showing the little relief that does exist
New Brunswick		1:662,000	58.8 × 59.7	French and English	Kilometres	Shaded relief	
Newfoundland and Labrador	Nfl. Labrador	1:1,200,000 1:3,000,000	58.4 × 58.4 27.9 × 35.6	English only	Kilometres	None	Forested areas shown in a blue stipple, which is strange considering that green is used on the map to show parks
Northwest Territories	SW All NWT	1:1,077,000 1:5,700,000	63.5 × 58.4 67.3 × 58.4	English only	Miles and kilometres	Hachures	
Nova Scotia		1:633,600	95.2 × 39.4	English only	Miles and kilometres	None	
Ontario	South North	1:800,000 1:1,600,000	86.4 × 68.6 70.0 × 70.0	English only	Kilometres	None	
PEI		1:253,440	38.1 × 43.2	English only	Miles	No relief to show	The smallest province, hence the largest scale road map
Quebec	East West	1:750,000 1:750,000	68.6 × 55.9 97.8 × 68.6	French only	Kilometres	None	
Saskatchewan		1:1,500,000	41.9 × 82.0	English only	Kilometres	None	
Yukon		1:1,500,000	41.9 × 82.0	English only	Kilometres and miles	Shaded relief	
Canada	East West	1:3,000,000 1:3,000,000	92.7 × 43.2 92.7 × 43.2	Separate English, French and German editions available	Kilometres	None	Newfoundland shown in inset

13 maps. Highways are classed according to width and whether they have hard or loose surfaces, the usual classification being: divided highway, main routes paved, main routes gravel, secondary routes paved, secondary routes gravel, and dirt (or unimproved) roads. The Northwest Territories and Labrador have fewer types, but even the Territories reserves a symbol for a paved road even though there are only 20 miles of such roadway in the region. The maps also have certain other common characteristics. All except the Canada Highway Map (the federal production) have indexes of cities, towns and villages; all have tables of mileages between important cities and towns; all employ from six to eight colours in the cartography; all except the Map of Canada, the Yukon and the Northwest Territories have plans of the important cities drawn at large scale in the margin; all except Prince Edward Island show distances in kilometres, though some (as indicated in the table below) have mileages as well; all are distributed folded, with information of tourist value; and all are supplied free to anyone requesting them through the provincial tourist organization.

Table 28 lists the important characteristics of each road map.

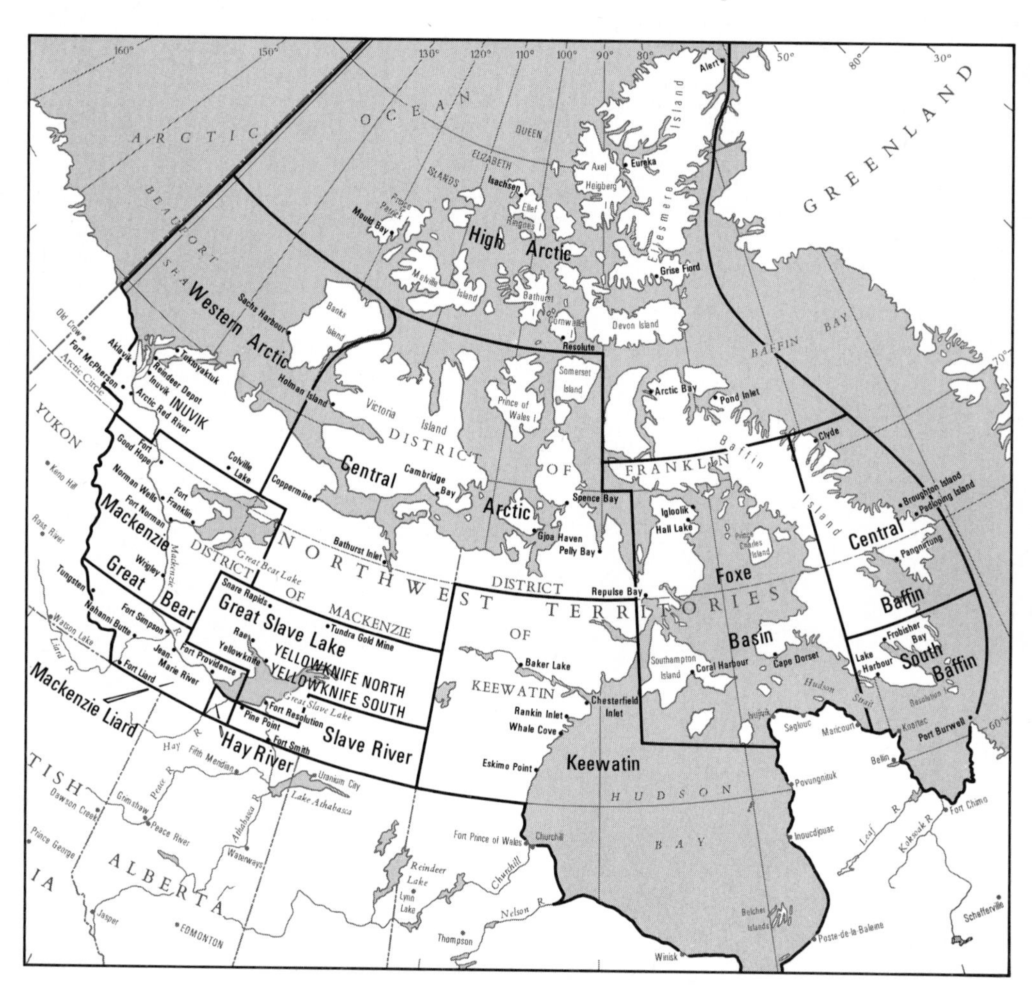

Figure 34 Electoral districts for the Northwest Territories

Electoral Ridings[3]

Canadians elect members to three levels of government: federal, provincial and municipal. The federal ridings number 282 and vary in size from the largest, which is half of the Northwest Territories, to the smallest, which consists of a few city blocks in downtown Toronto. Maps of each federal riding are prepared by the Surveys and Mapping Branch, and are amended as required by periodic redistributions which are made following the decennial census. Federal electoral maps are in two colours with the map detail shown in black and the district boundary in a red line. Various scales are used depending on the size of the riding. In addition to the maps of individual ridings, a map of the whole of Canada at 64 miles to the inch is available which shows all federal ridings (insets show the individual ridings in the larger cities). Also available are maps of federal ridings by provinces. All of the above maps are available through the Canada Map Office.

At one time provincial ridings largely followed county lines, but today this is the exception rather than the rule. Provincial electoral maps are prepared in, and are available from, the office of the Chief Electoral Officer in the parliamentary offices of each province. In the Yukon and Northwest Territories the territorial governments in Whitehorse and Yellowknife distribute the electoral maps for territorial council elections. These maps are drawn by the Surveys and Mapping Branch in Ottawa. Figure 34 shows the electoral districts in the Northwest Territories.

At the municipal level, electoral maps are prepared by local governments when they are required. They generally are needed in cities and towns, but in rural areas the township, parish or village boundaries are generally sufficiently well known and defined to make special maps unnecessary.

Table 29 The Numbers of Ridings (and of Electoral Maps) by Province and Territory 1980

	Federal ridings	Provincial (or Territorial) ridings
Newfoundland	7	51
Prince Edward Island	4	16
Nova Scotia	11	46
New Brunswick	10	58
Quebec	75	110
Ontario	95	125
Manitoba	14	57
Saskatchewan	14	61
Alberta	21	75
British Columbia	28	55
Yukon	1	12
Northwest Territories	2	15
	282	681

12. Hydrographic Charts

Hydrographic surveys and nautical charts could be considered the precursors of all Canadian mapping in general and thematic maps in particular simply because accurate soundings of Canadian waters were needed before any of the shores could be approached with safety. Some of Champlain's earliest surveys resulted in relatively large-scale charts of ports around the Atlantic coast and at least one of his small-scale maps, that of 1612, was expressly drawn for French navigators.[1] His small-scale maps in sequence indicate the increasing accuracy of the knowledge of the Great Lakes.[2] This combination of land and water surveys was continued much later (and with much more accuracy) by Samuel Holland and Joseph DesBarres in eastern Canada in the 1760s. From them James Cook learned the rudiments of marine surveying and charting and his 1763–7 surveys of the southern and western coasts of Newfoundland show his mastery of the art of hydrography.[3] From him George Vancouver learned the skills which were to extend the Pacific surveys during 1791 to 1795. Many of the early charts of the Arctic waters were produced by men in the same tradition. For example, Edward Parry who achieved outstanding triumph in northern navigation and surveying during his three Arctic expeditions, became Admiralty Hydrographer in 1825. Both the coasts of the Gulf of St Lawrence and those of the upper Great lakes were painstakingly detailed by another great hydrographer with the Admiralty Surveying Service—Henry W. Bayfield.[4] From 1816 to 1856 he surveyed practically the entire shoreline from Lake Superior to the Atlantic Ocean, produced detailed navigation charts and published the first *Sailing Directions for the Gulf and River of St Lawrence*. He is regarded as the pioneer of modern hydrography in Canada.

The British North America Act assigned the responsibility for navigation to the federal government in 1867. However, the length of coastline was at first confined to the Canadian portion of the Great Lakes, part of the Gulf of St Lawrence and the Atlantic coasts of New Brunswick and Nova Scotia. This was considerably extended in 1871 and 1873 when British Columbia and Prince Edward Island joined the federation, and even more so when the northern Arctic islands were added to Canada in 1880,[5] and when the federal government assumed the responsibility for hydrographic work in 1883. In that year, following the loss of the steamer *Asia* with some 120 lives 35 miles northwest of Parry Sound, the Georgian Bay Survey was established. The first chart to be issued under orders of the Government of the Dominion of Canada, entitled 'Cabot Head to Cape Smith and Entrance to Georgian Bay', was engraved by the British Admiralty and published in 1886.[6] Indigenous Canadian salt water surveys also began as the result of marine accidents. In 1890 the Canadian Pacific liner *Parthia* struck a shoal in Vancouver harbour and

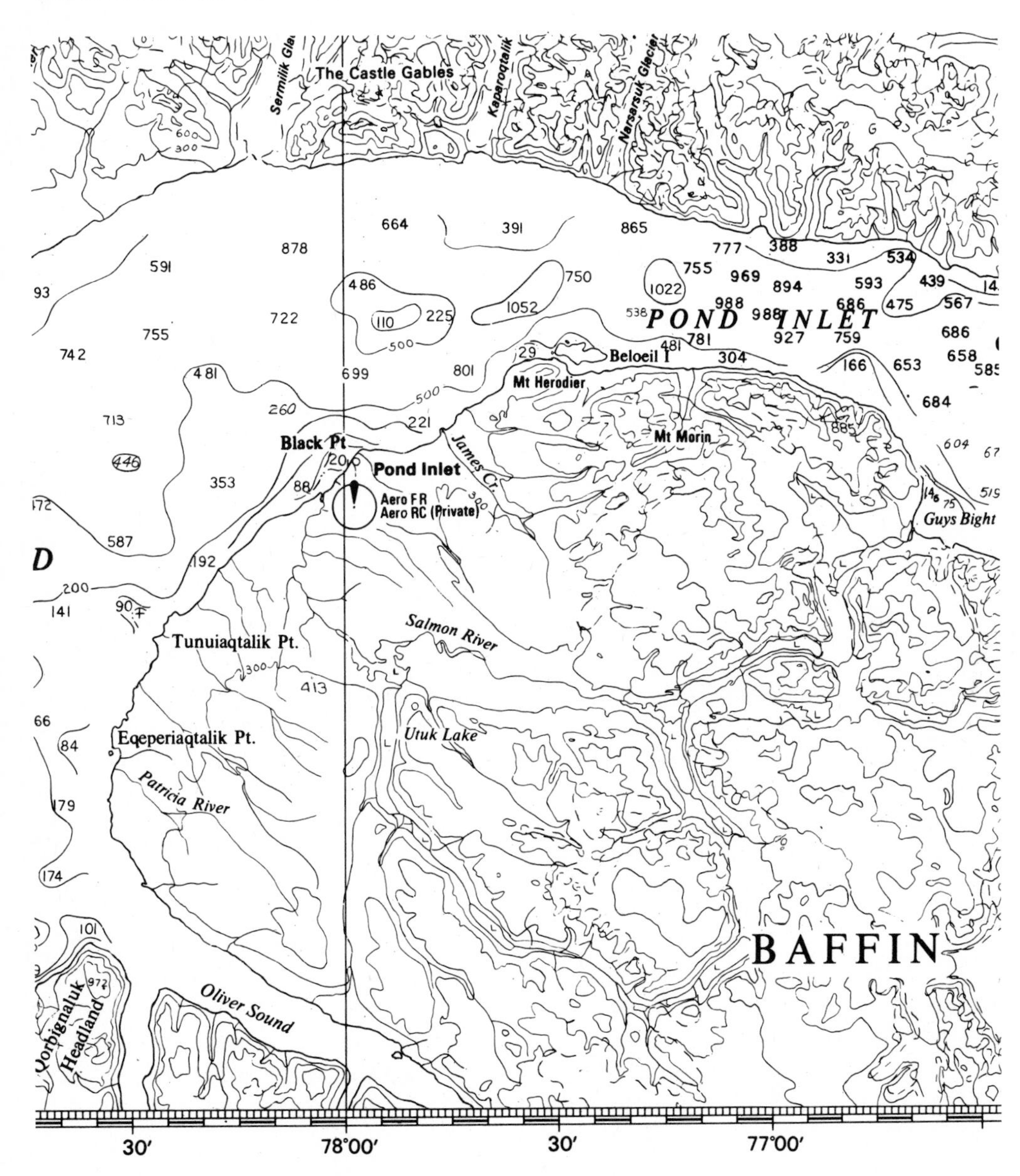

Figure 35 Section from a hydrographic chart

as a result a surveyor was detached from the Georgian Bay Survey to carry out hydrographic work in Burrard Inlet.[7] In 1904, again partly due to the loss of a vessel, this time in the St Lawrence River, the federal government amalgamated hydrographic and charting work within one department, 'with a view to systematizing and facilitating work in connection with Hydrographic Surveys' and, quite coincidentally, a few months later the British Admiralty formally requested that Canada assume full responsibility for hydrographic surveying of her own sea coasts. However, British Admiralty charting in Canadian waters did not completely end until 1911 (or 1932 if Newfoundland is considered). The earliest coastal charts produced under Canadian auspices were produced in 1907–8, engraved versions appearing in 1911. However the first chart to be published in Canada, 'Red River to Berens River', now Chart 6240, covering the southern part of Lake Winnipeg, was lithographed in 1904.

With the entry of Newfoundland into confederation in 1949, the Canadian Hydrographic Service became responsible for charting the longest coastline in the world, 151,489 statute miles, in addition to an extensive system of inland navigable rivers and lakes.

Conventional nautical charts are individually designed for use as instruments of navigation covering specific bodies of water and with an overlap to adjoining charts to permit positions to be readily transferred. Consequently, contrary to the uniform scale and size of topographic sheets, charts vary widely in both. Harbours and channels may be shown on a scale of 1:5000 to 1:20,000; for well-frequented coasts 1:75,000 is a typical scale, and there are smaller-scale charts in great variety. Most charts are on the Mercator projection and the small-scale charts are not provided with graphic scales owing to the great variation in scale with latitude. They are generally drawn on the true meridian.

Principal Symbols and Colours[8]

What distinguishes the general appearance of hydrographic charts from topographical maps is the dominance of the use of two colours, buff and blue; buff being used to cover all land areas and blue for all shallow water areas, with the two being combined to provide green for the areas between the low- and high-water lines. The symbols and abbreviations which appear on them are generally in accordance with the standard form prepared by the International Hydrographic Bureau in which all countries use the same sequence of numbers, although some items are added to suit Canadian conditions. A complete list of symbols and abbreviations is shown on Chart 1.

Soundings

Soundings are the most important information shown on a chart. The unit used is always stated in the title. Most charts show fathoms and feet in depths of less than 11 fathoms, and fathoms in deeper water. Some other charts use fathoms and fractions. Large-scale harbour plans, and charts of inland waters show depths in feet. Newer charts show metres and decimetres. The position of the sounding is at the centre of the space occupied. On most charts the density of the soundings gives a quick visual impression of depth with most soundings being shown in shallow water and very widely spaced in deep water.

Isobaths and Tints

Isobaths are generally drawn for the depths 1, 2, 3, 5, 6, 10, 20, 50, 100, 500 and 1000 fathoms, and each has its own symbol. A blue tint is added between the low-water line and the 3-fathom line, and a half-blue tint from the 3-fathom line to the 6-fathom line. The limiting depth of the tint is indicated in the chart title. On some large-scale plans or charts of inland waters the isobaths are drawn in feet but reflect the same intervals. On all new charts issued after 1977, together with earlier international charts and charts of Arctic waters, depths and isobaths are now shown in metres. A new format is now being introduced in which the isobaths are shown in solid blue lines for 2, 5 and 10 metres, then at 10-metre intervals to 100, at 20-metre intervals between 100 and 200, and at 100-metre intervals beyond 200 metres. On the new format far fewer soundings are shown, but the isobath labels are staggered to provide a balanced display of depth information throughout the chart.

Heights

On charts of tidal waters, heights are given in feet above high water. On charts of non-tidal waters and for drying banks or drying rocks, heights are expressed in feet above the chart datum.

Coast Features and Dangers

Particular attention is paid to these items. The following coastal characteristics are each indicated by an appropriate conventional symbol:

Steep coast
Cliffy coastline
Sandhills or dunes
Sandy shore
Surveyed coastline, which is identical to the high-water line
Low-water line

In addition the following foreshore characteristics are indicated for areas which are bare at chart datum:

Mud banks
Sand
Stones, shingle or gravel
Rocky ledges and isolated rocks
Sand and mud
Sand and gravel or stones

Within the limiting danger line, particular dangers are identified by means of different symbols and abbreviations. Four of these are concerned with different kinds of wrecks and six with different kinds of rock dangers. The quality of the bottom, such as mud, shingle or pebbles, is designated by an appropriate abbreviation.

Aids to Navigation

Whereas soundings are shown on charts to provide depth information at any particular location, aids to navigation are shown to enable the navigator to determine the position of his vessel in relation to navigable channels and shoals that constitute a danger to shipping. Lights, buoys and beacons, radio and radar stations and fog signals established for this purpose, are indicated using an extensive variety of symbols and their exact characteristics described by abbreviations. They are usually printed in black but are sometimes in nautical purple (especially for radio and radar stations)—the third colour used on charts.

In addition to these aids, lines of position for electronic positioning systems are shown on coastal and offshore charts. These systems include Decca, LORAN-A and LORAN C and enable navigators to determine a ship's position by using receivers on bands that pick up signals from stations on shore.

Land Features

The symbols on the land surface are the same as on topographic sheets although emphasis is placed on the conspicuous features which can be used by the mariner in fixing his position. Thus conspicuous hills, towers, chimneys, church spires, etc. are identified because they can be easily seen from ships off the coast.

On some charts where no topographic relief information is available the land is left completely blank apart from information pertaining directly to navigation, such as lighthouses, landmarks and fresh water sources.

Notices to Mariners

To ensure the safety of navigation it is essential that charts be up-to-date at all times. To permit this the federal Ministry of Transport publishes weekly notices to mariners. These show the corrections that must be made by hand. The last weekly notice to be incorporated into a chart is indicated by a rubber stamp with the date of issue. The numbers of the specific notices incorporated are shown in the lower left-hand corner. It should also be noted that a new edition of a chart supersedes the previous edition which must no longer be used for navigation.

Small Craft Charts[9]

The number of small boat operators increased dramatically after the Second World War, and by 1960 they were responsible for about 60 per cent of Canadian chart sales. As standard-sized nautical charts are not suitable for use in small craft, it was decided to produce specially designed small-craft charts for their needs. They were inaugurated in 1946 with a set of accordion-folded strip charts covering the sheltered passage from Parry Sound to Byng Inlet on the east coast of Georgian Bay. They also differ from the standard charts in that more use is made of colour, with the danger isobaths of three to six feet being emphasized. Mileages, recommended tracks and marina locations are shown in colour, and access highways and launching ramps are shown. Legends used on a particular chart are printed on the chart. Where the chart portrays a long narrow waterway the chart is usually oriented along the waterway rather than on true north.

Chart Indexing

Canadian hydrographic charts also differ from topographical maps in their index numbering system. The hydrographic charts were initially assigned a series of numbers ranging from 1 to 999. In the early 1940s a four-number system was introduced to permit the expansion of the number of charts on issue and to provide for the assignment of new numbers to charts replacing old charts being withdrawn.

The four-number system ranged from 1 to 9199 with a block of numbers assigned to coastline areas as follows:

Miscellaneous non-navigation charts	1– 999
St Lawrence River and tributaries	1200–1999
Great Lakes and tributaries	2000–2999
Pacific Coast and Pacific drainage	3000–3999
Atlantic Coast	4000–4999
Hudson Bay and Strait	5000–5999
Inland waters excluding the Great Lakes	6000–6999
Arctic	7000–7999
Fisheries charts	8000–8999

The above major divisions are sub-divided with blocks of numbers assigned to specific coastline areas within the major division.

The limits and scales of charts are shown in four chart catalogues:

1 Atlantic Coast
2 Pacific Coast
3 Great Lakes
4 Arctic

These catalogues and individual charts are obtainable from the Hydrographic Chart Distribution Office, Fisheries and Environment Canada, P.O. Box 8080, Ottawa, Ontario K1G 3H6.

13. Important Provincial Map Series

All provinces of Canada publish maps and some are very active in this field. The total number of provincial map series is well over 400, so this review will consider only those that will be most useful to researchers and students. The various series will be grouped by provinces under the headings: topographical, geological and other thematic maps. In general, series smaller than 1:250,000 and larger than 1:2000 have not been included.[1]

Much of the provincial mapping is reproduced by a photochemical system which provides maps consisting of black or dark blue lines on white paper. The original copy is a film positive which can be produced by process photography from scribed line work and peeled open-window negatives for areas, such as lakes. Screens can be used to provide grey lines or areas. A number of different types of equipment are available to do this work, but to avoid the use of trade names, the maps so produced will be called white-prints. With modern equipment the line work is almost as sharp as that produced by lithography. Photomaps can also be reproduced by white-print machines, and again they are not much inferior to the lithographed equivalent.

One additional general comment should be made to draw attention to the recent entry of provincial governments into the business of topographic mapping. Fifteen years ago only British Columbia produced such maps, but today all provinces are either active in this work or are making plans. One of the more popular scales is 1:20,000 which of course is in direct competition with the dormant federal 1:25,000 scale.

British Columbia

Topographic Maps

British Columbia has a long tradition in topographic mapping and for many years cooperated with the federal government in the production of 1:50,000 sheets of the federal series.

The sheets done by the province were surveyed, plotted and carried through the fair drawing stage before being shipped to Ottawa for printing. Unfortunately, because of the pressure of other work this arrangement was terminated in 1968. The province continues to publish lithographed topographic maps at scales of 1:250,000 and 1:100,000. (The latter scale will gradually replace the excellent 1:125,000 series, and will have identical specifications except for the scale.) These series differ from the federal 1:25,000 and 1:125,000 maps by having much more cadastral information (in fact, they are called land status maps). They also have a more detailed road classification but do not show the forest cover.

The province also publishes a range of large-scale topographic-cadastral maps at scales of 1:2000, 1:2500, 1:5000, 1:10,000 and 1:20,000 which are reproduced by white-print. A number of photomaps formerly at the scale of 1:31,680 but more recently at 1:2500, 1:10,000, 1:20,000 and 1:40,000 are also published. This work is done in the Surveys and Mapping Branch of the Ministry of the Environment in Victoria.

Geological and Related Maps

Geological mapping in British Columbia is both voluminous and of a high standard. The most detailed maps are those accompanying reports of the various geological explorations that have been carried out by provincial geologists, but these are individual maps—not part of a series. An important geological series is the preliminary Geological Map series, published on a range of scales from 1:12,000 to 1:250,000, giving in monochrome display the outline geology of a number of the important geological features of the province. Another important mineral series is the revised Mineral Inventory Map series which covers the whole province at the scale of 1:125,000 (older sheets at 1:126,720 or 1:250,000). These white-print sheets serve as a data base for filing information on mineral claims, and are updated frequently.

Another white-print map series is the Assessment Report Index series. The sheets of this series record the location of filed reports on mineral deposits and claims, and the assessment work being done to keep the claims in force. They are published at the scales of 1:126,720 or 1:250,000.

An important series at 1:250,000 is the Mineral Deposit–Land Use Map series which is published by white-print. These sheets address the rather 'delicate' subject of mining and petroleum production in the context of other land use. They complement the maps on renewable resources produced by the Resource and Analysis Branch of the Ministry of Environment, which will be mentioned shortly. The provincial agency concerned with geology, mining and petroleum production is the Department of Mines and Petroleum Resources which has its headquarters in Victoria.

Other Thematic Maps

The British Columbia Forest Service produces and maintains a series of white-print forest inventory maps showing the stands of merchantable species on which the tree volume, ages, conditions and status of the stands are given. Other maps show forest district boundaries, the location of public sustained-yield units, watershed reserves, timber sales, tree farm licenses, research plots and other relevant data. Forestry is a very important industry in British Columbia, and the forestry map production is a reflection of this activity.

As has been mentioned, The Resource and Analysis Branch of the Ministry of the Environment produces maps relating to the inventory of renewable resources. Some of the lithographed maps, which show soils capability for agriculture, forestry, recreation and wildlife, are published in cooperation with the federal Department of Environment. Other maps illustrate various biophysical characteristics such as soils, climate, vegetation, present land use, etc. The task of the Branch is to prevent excesses in the exploitation of the province's natural resources and to ensure a continuation of the resource industries for all time. The maps they produce are basic tools in this endeavour.

Alberta

The Alberta government is at the time of writing (1979) reorganizing its whole mapping system to provide a metric range of scales with cartographic specifications conforming to present-day requirements. The overall plan is to produce maps at the following range of scales: 1:1000, 1:5000, 1:10,000, 1:20,000, 1:50,000, 1:100,000 and 1:250,000. These scales are designed to cover both the topographic and thematic mapping needs of the province without duplicating the federal government's mapping at 1:50,000 and 1:250,000.

Two provincial departments are involved in this work. The Alberta Ministry of Transportation will be responsible for the three largest scales, which will be cadastral in nature, and the 1:250,000 series which will be a frequently updated large-scale road map. The other scales will be produced by the Alberta Ministry of Energy and Natural Resources. As the new programme will take a number of years to replace existing mapping, this report will, of necessity, cover the old and the new.

Topographical

In 1975 Alberta published topographic maps at a scale of 1:25,000 covering the Alberta Oil Sands area (quadrangles 74 D and E and 84 A and H of the National Topographic System). Each sheet was published in two styles: a contoured orthophoto map and a monochrome line map. These maps were produced by the Surveys and Mapping Branch of the Alberta Ministry of Transportation, but there is no plan to expand or maintain this scale. The 1:20,000 topographic series being planned by the Alberta Ministry of Energy and Natural Resources will replace this series.

The 'Alberta Topographic Series' at the scale of 1:126,720, published by the Alberta Ministry of Energy and Natural Resources will be converted to the 1:100,000 scale. This series is lithographed in six colours: black (man-made features except roads), brown (contours), blue (hydrography), green (forest cover), red (roads) and yellow (park boundaries and the built-up areas of cities). Only 10 sheets have been published at 1:126,720 but it is expected that more resources will be put into this series at the new scale, and that production will be linked to the 1:100,000 Forest Cover series which will be described later.

Geological

Provincial geological mapping in Alberta is done by the Geological Division of the Alberta Research Council, Edmonton. In general, the topographic maps of the National Topographic System are used as base maps, but occasionally special maps are designed to fit the geological feature being reported. The following are the more important provincial geological series.

Bedrock Topography
 2 sheets at 1:50,000
 12 sheets at 1:250,000
Surficial Geology
 21 sheets at 1 inch to 1 mile or 1:50,000
 10 sheets at 1:250,000
Glacial Geology
 8 sheets at 1 inch to 1 mile
 1 sheet at 1 inch to 4 miles
 1 sheet at 1 inch to 8 miles
Hydrogeology
 3 sheets at 1:125,000
 14 sheets at 1:250,000
 2 sheets at 1:500,000
Economic Geology
 22 sheets at various scales from 1 inch to 2000 feet, to 1 inch to 32 miles
Urban Geology of Edmonton
 16 sheets at 1:50,000

As these series are revised or extended they will be made to conform to one of the metric scales mentioned in the introduction to this section.

Other Thematic Maps

Cadastral mapping by Alberta Transportation will continue at the scales of 1:1000, 1:5000 and 1:10,000. At present over 400 sheets have been published at 1:5000. The 1:1000 series will replace the 1:2500 series which is dormant in three experimental sheets. One experimental sheet at 1:10,000 has been published (without contours), but the future of this series scale is not clear.

The new 1:250,000 series being undertaken by the Alberta Ministry of Transportation will replace the 1:253,440 coloured series previously produced by the Alberta Ministry of Energy and Natural Resources under the title of 'Provincial Access Series'. The new series promises to be most useful, and deserves a short description here. It will be published in two styles, in monochrome for use as a base map for thematic mapping, and in four colours for general use. The colours used will be black for populated places, railways, pipelines, power lines, the DLS survey pattern (where it does not coincide with roads) and symbols for airports, helipads, etc. Red is used for paved and gravel roads and for forestry stations and towers, blue is used for hydrography, and brown is used for unimproved roads, truck trails and seismic lines. Contours and the vegetation and marsh symbols are not used. The philosophy behind this series is that the federal maps at the same scale are revised so infrequently (generally 10 years in most of Alberta) that the access information is, generally speaking, seriously out-of-date. The provincial authorities plan to revise the maps on a two-year cycle.

The Forestry Branch of the Alberta Ministry of Energy and Natural Resources has published over the years a white-print series known originally as the Provincial Record series, but more recently as the Forest Cover series. These sheets are published at 1:63,360 and show the density, species and maturity of timber stands. Background information shows the DLS cadastral system, access roads, seismic lines, pipelines and other man-made features. Coverage of the province is complete except for the treeless southeast corner of the province and the federally owned Wood Buffalo Park on the north boundary of the province. Virtually the same information is published at 1:126,720 in four-colour lithographed maps. The colours are used as follows:

Black	Border and legend detail, black topography, black names, survey detail; also contour information is screened grey and is part of the black plate.
Blue	All hydrography and blue names.
Red	All roads and seismic lines. All usable airstrip and other forestry installations.
Green	Forest cover—a broad inventory of forest stands identified by symbolization for density, height and species.

The future plans see the 1:63,360 published as a monochrome 1:50,000 map and the 1:126,720 series enlarged to 1:100,000. The colour plates of the latter series will be drawn so that they can be used as required for the 1:100,000 topographic map mentioned earlier.

Saskatchewan

Topographical

At the time of writing (July 1979) Saskatchewan has not published topographic maps, but plans are underway to produce large-scale (probably 1:5000) coverage of the cities of Regina, Saskatoon and Prince Albert. This activity is being organized within a new establishment called the Central Survey and Mapping Agency in the Department of Highways and Transportation in Regina.

Geological

Geological mapping in Saskatchewan is done by the Department of Mineral Resources. Geologically the province is divided into two distinct regions, the Precambrian Shield generally to the north, and younger sedimentary formations to the south. The line of separation runs roughly from a point at latitude 57° on the west boundary of the province southeast through Frobisher Lake, Lac la Ronge to the town of Flin Flon on the Manitoba border. The geological mapping in the south is mainly in support of oil exploration, but because the directly observable geology can only be obtained from rather widely spaced drill holes, this mapping tends to be at small scales—generally the whole of a particular geological area being shown on one sheet. This requires the use of odd scales and sheet lines rather than the 'map series' approach. The Saskatchewan potash deposits have been studied and mapped. In the Precambrian Shield area geological mapping is carried out in support of the exploration of mineral resources, and much larger scales are used. Now that the federal 1:50,000 sheets are available, these are used as base maps for the provincial series. In the past, Saskatchewan has used scales of 1:31,680, 1:63,360, 1:126,720 and 1:253,440 and these sheets are still available for many areas, but in the future scales of 1:20,000, 1:50,000, 1:100,000 and 1:200,000 will be used.

Other Thematic Maps

There are three categories of forest inventory maps published by the Forestry Branch of the Department of Tourism and Renewable Resources. These are:

Forest inventory maintenance maps—scale 1:15,840
Forest inventory reconnaissance maps—scale 1:25,000
New forest inventory maps—scale 1:12,500

Township plans are published by the Department of Tourism and Renewable Resources, Lands and Surveys Branch, and are at the scale of 1:31,680.

Maps of the rural municipalities are published by the Department of Municipal Affairs at a scale of 1:126,720.

Manitoba

Topographical

The province of Manitoba has recently embarked on the production of white-print topographical series on a scale of 1:20,000, with a contour interval of 5 or 10 metres depending on the terrain. The first sheets cover the areas around the settlements in the northern part of the province, but in time this coverage may be extended. Each sheet is also published as a contoured photomap.

Geological

Manitoba, like Saskatchewan, has two geological regions, the Precambrian Shield to the north and Palaeozoic formations to the south. The dividing line runs from Flin Flon at the west boundary, through Lake Winnipeg to the southeast corner of the province. The Mineral Resources Branch of the Department of Mines, Natural Resources and Environmental Management publishes mineral claim sheets at scales of 1:31,680 and 1:50,000 in areas of the north of the province where there is prospecting or mining activity.

Other Thematic Maps

Cadastral plans of townships and rural municipalities are published by the Surveys and Mapping Branch of the Department of Mines, Natural Resources and Environment

Management. These are restricted to the southern part of the province. Forest inventory mapping is done at three scales: 1:63,360 covering the whole of the forested area of the province; 1:31,680 covering the northern portion of the woodlands; and 1:15,840 covering the southern portion of the woodlands.

A useful series of lake depth charts is published to support both game fishing and commercial fishing. These are drawn at a multitude of scales, each scale being chosen to fit the lake on 8½ inch by 11 inch paper.

Ontario

Topographical

In 1978 the Ontario government's Surveys and Mapping Branch (Ministry of Natural Resources) embarked upon the publication of the Ontario Base Map series, an extensive system of white-print topographic maps on three scales, 1:2000, 1:10,000 and 1:20,000. The contour interval for the 1:2000 scale is one metre, for the 1:10,000 scale, five metres, and for the 1:20,000 scale, 5 metres or 10 metres depending on the roughness of the terrain. They are all drawn on the UTM projection, show spot heights, the forest cover and man-made features, especially roads and buildings. (As a parenthetical observation, it might be of interest to mention that in forested areas beaver dams are shown. This is a topographic feature of frequent occurrence and considerable permanence not previously identified on any Canadian topographic series.)

The coverage planned for this system of maps is as follows. Northern Ontario will be mapped at 1:20,000. The southern limit of this area is not rigorously defined, but in general it follows a line passing immediately to the north of North Bay and Sudbury, and then along the north shore of Lake Superior to the international boundary. Within this region the Hudson Bay Lowland will be depicted on 1:20,000 orthophotomaps because of the swampy, featureless nature of the country. The rest of northern Ontario will be covered by 1:20,000 line maps except for a few settled areas where 1:10,000 maps may be drawn. Southern Ontario will be mapped at 1:10,000 with heavily urbanized areas also covered at 1:2000.

Since the middle of the 1960s Ontario has published what is called a topographical series, but they are actually planimetric maps in colour principally showing alienated lands. Man-made features are shown by means of the conventional symbols usually used for the NTS series. Indian lands and areas of alienated surface rights are shown in buff colour and the rest of the land surface is coloured green. Water areas are in blue. The series is on a scale of two miles to one inch (1:126,720) and the projection used is the Transverse Mercator. The maps cover principally the area of northern Ontario between 46°N and 50°N west of 82°W and the sheet lines are in accordance with those of the NTS 1:125,000 series. They are produced by the Ministry of Natural Resources, Toronto.

Geological and Related Themes

Provincial geological mapping in Ontario is done by the Ministry of Natural Resources. A geological map of the whole province is available in five sheets each on a scale of

1:1,012,760 (16 miles to 1 inch). In addition over 500 maps are available on larger scales usually accompanied by a report. The maps are almost all planimetric, but the scales are varied to fit the geological features of the area. The polyconic, Lambert conformal conic or the Transverse Mercator projections are used.

An extensive series of planimetric maps showing oil and gas wells, generally on a scale of 1:31,680, but often on a scale of 1:15,840, covers most of the southwestern part of the province.

A mineral deposit series shows Ontario mineral potential by means of black overprint on a monochrome green topographical base. The maps in this series are on a scale of four miles to one inch (1:253,440). Seven have been published so far covering the southern part of the Canadian Shield.

A quite recent series published by the Ministry of the Environment is devoted to the hydrology of drainage basins and related matters. The scale varies with the drainage basin, but generally ranges from 1:100,000 to 1:500,000. Most of the maps are planimetric on the Transverse Mercator projection. For the Moira River drainage basin, for example, the scale used is generally 1:200,000 and maps are available for the basin on (a) physiography and sub-basins, (b) hydrometric stations, (c) bedrock geology and topography, (d) generalized surficial geology, (e) locations of water wells, (f) overburden aquifers and estimate yield, and (g) groundwater availability and daily streamflow exceeded 90% of the time.

A. Ground Water Probability series of maps is published by the Ontario Water Resources Commission, and in 1977 it covered the counties of Essex, Lambton, Kent, Elgin, Haldimand and Brant on a scale of 1:100,000 on the Transverse Mercator projection. The maps indicate the probable water yield from wells, the probable depths of the water-yielding formation, and areas where shallow wells and sand points may be developed; this and associated water data are in colour. The base maps are not contoured but show concession and lot lines and topographic symbols for man-made features in black.

Other Thematic Maps

The government of Ontario publishes a wide range of other thematic maps, the principal series being as follows:

COUNTY, DISTRICT AND REGIONAL MUNICIPALITY SERIES. White-print maps, produced as monochrome black line paper prints for individual counties, districts and regional municipalities are published, the maps of the larger administrative units being in sections on two or more sheets. Township, lot and concession lines are shown. The scales used are either 1:63,360 or 1:126,720.

Maps produced by lithography in colour are available from the Ministry of Transport and Communications for counties or combinations of counties, regional municipalities and districts on a scale of 1:253,440 or 1:250,000. This is one of the oldest of the provincial series and is clearly a successor to the federal Chief Geographer's series described in Chapter 10. It shows all roads, township boundaries, and concession lines with farm lots numbered at intervals.

MUNICIPAL BOUNDARY HISTORY SERIES. This series has been published for several of the regional municipalities of Ontario. It is really a series of sets, as the number of sheets for a municipality varies from 3 to 7. Intended to show the changes that have occurred in the municipal boundaries within the parent municipality, they are printed in black and white and shades or intensities of one other colour (the 'other colour' being different for each regional municipality). The scales for the municipality as a whole vary from four to five miles to one inch.

FOREST RESOURCES INVENTORY MAPS. These cover the whole province as far north as 52°N and in the western portion above 53°N. These are available as ozalid prints and show general topography without contours. They are drawn from semi-controlled air photograph mosaics on a scale of approximately 1:15,840. In time these will be converted to the 1:20,000 scale.

LAND INVENTORY MAPS. Ontario has published a series of land inventory maps which differ in style and approach from the maps derived from the federal Land Capability series described in Chapter 11. All the maps in the Ontario series consist of information overprinted in black on the NTS 1:250,000 series.

Fundamental to this series are the maps of land classification on which landscape units are identified by local names in purple and their boundaries are outlined in purple lines. Landscape units are a grouping of one or more land units and, in many cases, one or more water units which are related to one another because of a similarity of physical features, a dominance of cultural features such as an agricultural community, or a forest plantation area, or a distinctive pattern of land and water features.

Land units are identified by numerical symbols and their boundaries are outlined in green lines. Land units are areas of land having a distinctive pattern of physical features such as relief, soil materials, depth of soil materials and soil moisture. Thus 62-4 designates the fourth land unit within a specific landscape unit which belongs to Class 62.

These sheets cover almost all of Ontario south of 52°N.

For each land unit delineated and described on the land classification maps an evaluation of the general timber use capability pattern is indicated within the boundaries of each land unit using a figure and letter code on separate map sheets. These timber use capability maps cover all of Ontario south of 48°N, with some for areas north of this.

Land capability for recreation has been similarly mapped, but most of the 1116 maps are available only as ozalid prints on a scale of 1:50,000. There are also 910 maps of land capability to produce wildlife available as ozalid prints on the same scale, and there are 1350 maps on the recreation capability for bathing and lodging along shorelands on a scale of 1:15,840.

FISHING MAPS. This series consists of over 400 black-and-white maps lithographed on water resistant paper. There is one map sheet for each surveyed lake in Ontario, unless the lake is very large in which case two, or even three, sheets are used to cover it completely. Each map shows the shoreline of the lake, and bathymetric contours are sometimes supplemented with spot depths. The contour interval varies from map to map, but is generally between 10 and 20 feet. The scale also varies, but is within large-scale limits, being of the order of 1:12,672 to 1:31,680. Access routes to each lake are also shown. No latitudes or longitudes appear on the maps themselves, but such data is printed along with other information in a panel in the map margin.

Quebec

Topographical

Quebec has an active topographic mapping section within the Ministère des Terres et Forêts which publishes contoured maps at scales of 1:1000, 1:2000 and 1:20,000. The three series are published in white-print, but because the reproduction material is scribed very sharp linework is produced. Most of the symbolization for the 1:20,000 series is the same as on the Ontario mapping of the same scale, described earlier, but minor variations can be noted. For example, depression contours are indicated by tiny arrows pointing downhill, rather than the normal contour ticks, and a special symbol resembling a series of small check marks is used to depict brush (broussaille).

Quebec's large-scale topographic mapping programme is a development from the 1:15,840 base map series published in the early 1950s. Originally these maps were required to map forest inventories, but gradually other uses and users were found, and in 1965 the scale was enlarged to 1:20,000 to better serve the requirements of town planning, resource development and cadastral work. In 1973 a 10-year programme was launched to complete the coverage of the province south of the 52nd parallel at the 1:20,000 scale. Within this programme was an accommodation to map certain highly developed regions at 1:10,000 and urban and suburban areas at 1:1000 and 1:2000. This programme is progressing on schedule.

Geological

The Ministère des Richesses Naturelles publishes a number of geological and aeromagnetic survey maps at various scales ranging from 1:12,000 to 1:63,360. Most of these are in white-print form. The aeromagnetic maps are published on a photomap base. In the future the provincial 1:20,000 and federal 1:50,000 maps will be used as base maps.

Other Thematic Mapping

A small-scale cadastral series is published at 1:200,000 showing counties, parishes and farm lots on an outline planimetric base. This series covers the portion of Quebec south of latitude 47°, along both shores of the St Lawrence and Gaspé Peninsula.

Forestry maps are published at scales of 1:20,000 (forest inventory), 1:50,000 (a reduction of the 1:20,000 series) and 1:125,000 (a synthesis of the inventory information, ground accessibility, etc.). In addition the Ministère des Transports publishes very detailed planimetric maps showing the various means of ground transportation. These sheets are lithographed in two colours (black and blue) and are published at scales of 1:50,000 and 1:125,000. It should be noted that this coverage at 1:125,000 is the only up-to-date series that fills the gap in the federal mapping of Quebec between the 1:50,000 and the 1:250,000 scales.

An interesting little series published by the Ministère des Transports is the 'Sentiers de Motoneige', a 1:190,080 series lithographed in colour that is very popular with snowmobilers.

Land use maps are published by the provincial Department of Agriculture at a scale of 1:50,000 (white-print), and by 1979 all of southern Quebec will be covered.

A series of quarter township maps (i.e. each township is covered by four sheets) is published at the 1:12,000 scale.

The Maritime Provinces

Topographical

The provinces of New Brunswick, Nova Scotia and Prince Edward Island must be considered together because in recent years they have pooled their resources in two mapping agencies, the LRIS (Land Registration and Information Service) and MRMS (Maritime Resources Management Service). The former, with headquarters in Fredericton, New Brunswick and offices in Halifax, Nova Scotia, and Summerside, Prince Edward Island, produces large-scale contoured photomaps at scales of 1:10,000 and 1:5000 and large-scale line maps at 1 inch to 100 feet, 200 feet and 400 feet. Cadastral maps using both the topographic and the photomaps are also produced. These maps show parcel boundaries and a parcel identifier for which computerized ownership information is available. The MRMS, situated at Amherst, Nova Scotia, produces mainly thematic maps, but on request publishes 1:15,840 and 1:10,000 planimetric maps derived from federal 1:50,000 maps with some additional detail derived from air photography. It is also producing two series of special topographic maps of Nova Scotia at 1:250,000 and 1:125,000. These differ from the federal series of the same scales by not having the green forest cover symbol and by showing all road classes with either red or black symbols. In this way the map can be printed with four colours: black for man-made features, blue hydrography, brown contours and red for the paved road classes. The maps are small in format and square in shape being cut along UTM grid lines and covering 50 kilometres in both directions. Forty-six sheets cover the province, and at the 1:250,000 scale are published in a booklet. The 1:125,000 maps, which are simply an enlargement of the 1:250,000 sheets, are published as separate maps. In the 1:250,000 scale a second series covering entire counties is in progress.

MRMS has produced one pilot topographic line map at 1:20,000 scale with the topography derived from the LRIS photomaps at 1:10,000, but it is too early to say that this will develop into a topographic line map series similar to those of Ontario and Quebec.

Geological

Because of the extensive work done in Nova Scotia and Prince Edward Island by the Geological Survey of Canada in the past, very little detailed geological mapping is taking place in these provinces at present. Some small-scale geological maps for both provinces have been produced by the MRMS. The Mineral Resources Branch of the Department of Natural Resources of New Brunswick is active in producing a white-print geological series at 1:15,840.

Other Thematic Mapping

The province of Nova Scotia is producing land use maps at 1:15,840. These are based either on the photomaps of LRIS or on the outline maps of MRMS.

Newfoundland and Labrador

Topographical

Newfoundland and Labrador Hydro are producing, by contract, white-print topographic maps of certain rivers with hydro-electric potential. These are drawn at a scale of 1:10,000 with a five-metre contour interval. Only the immediate watershed of these rivers is shown.

Geological

The Department of Mines and Energy produces geological maps at scales of 1:50,000 and 1:100,000. The NTS 1:50,000 series is used as the base for both series with the geological depiction being added by provincial geologists. Both series are lithographed in black and white.

Other Thematic Maps

The Department of Forestry and Agriculture publishes an extensive series of 1:12,500 forest industry maps in white-print. The bases for these maps are usually the NTS 1:50,000, which have been enlarged to scale with only the essential planimetric detail being shown. On these bases cartographers outline and identify timber stands from data on aerial photographs which have been marked up by foresters in the field.

Also in white-print are two series at the 1:1250 and 1:2500 scales published by the Department of Municipal Affairs and Housing. This series shows municipal services (roads, power lines, drainage, etc.) in cities, towns and their suburbs.

14. Atlases

Atlases of Canada

Reference has already been made to the early atlases of the counties of the eastern provinces, all of which were produced by commercial firms. An atlas which covered all of Canada was also published at this time by Walker and Miles who brought out four of the Ontario county atlases a year or two after their *New Standard Atlas of the Dominion of Canada* appeared in 1875. This must be regarded as the first 'commercial atlas' of Canada.

The first atlas to be produced by a Canadian government was the *Electoral Atlas of the Dominion of Canada*[1] which was published in Ottawa in 1895 and which followed the pattern of a similar publication for New South Wales, Australia. The Canadian atlas covered the populated portions of the provinces of British Columbia, Manitoba, Quebec, Ontario, New Brunswick, Nova Scotia and Prince Edward Island in 202 maps. Each sheet depicts an electoral district with its political subdivisions, railways, rivers and streams. The maps do not show roads but they do show postal routes as the postal maps were used as a base. The scales of the postal maps is also retained—72 miles and 24 miles to 1 inch (1:4,561,920 and 1;1,520,640) in the case of British Columbia; 12 or 18 miles to 1 inch (1:760,320 or 1:1,140,480) for Manitoba; 4 miles to 1 inch (1:253,440) for Ontario and 6 miles to 1 inch (1:380,160) for the remaining provinces. The Canadian atlas was produced by the Department of the Secretary of State, as was the electoral atlas published in 1906. This atlas of 211 maps was slightly larger than the 1895 version owing to the need to include the electoral districts of the Northwest Territories (on scales ranging from 24 to 35 to 40 miles to 1 inch) and the Yukon (on a scale of 75 miles to 1 inch). But by this time the position of Geographer had been created in the Department of the Interior and the drawing of the maps appears to have been entrusted to his office. The introduction to the atlas simply states that the maps were prepared by 'a skilled officer of the staff of the Department of the Interior'. There are clearly many revisions to the base maps including the omission of the postal routes. The even larger 1915 edition of the electoral atlas was entirely a publication of the office of the Chief Geographer (as the Geographer had become known by then).

The First and Second Editions of the National Atlas

It was also the office of the Geographer that produced the *Atlas of Canada* in 1906—a collection of thematic maps which is regarded as the first 'national' atlas of Canada.[2] In its formative stages in 1901–1902 it was referred to as an *Economic Atlas* intended to show 'in a general way the enormous, though largely undeveloped mineral, agricultural and forest resources of the country and . . . the great possibilities of development'. Canada was undergoing rapid development at the turn of the century and immigration was being actively encouraged especially in the Prairies. This was of particular concern to the Department of the Interior under whose *aegis* the atlas was produced, and there is some evidence to suggest that the *Economic Atlas* was originally planned as an atlas of western Canada. Furthermore 1906 would have been an ideal year to publish such a work as the provinces of Alberta and Saskatchewan had been created in 1905. The only other country to publish a national atlas up to this time had been Finland in 1899. In 1895, at the Sixth International Geographical Congress in London, England, the Geographical Society of Finland had exhibited a number of maps which ultimately were included in the atlas, and it is known that three Canadians attended the Congress—Sir Charles Tupper, then Canadian High Commissioner to the United Kingdom, the Hon. G. W. Allen of Toronto and Dr Henry M. Ami of Ottawa.[3] Whether they saw or were impressed by the Finnish maps, which were to result in an entirely new type of atlas, is not known. The Geographer of the Department of the Interior was one of the Canadians in attendance at the Eighth International Geographical Congress in 1904, but by that time the compilation of the *Economic Atlas* was well under way. His attendance may, however, have resulted in the change of title.

Unlike the electoral atlases, the majority of the maps in the 1906 *Atlas of Canada*[4] are on small scales and embrace a number of different topics. They deal with relief, geology, communications, forests, climate, population origin and density, drainage basins, routes of explorers and major political boundaries. Plans of the principal cities are also included. Three scales were used for almost all of the maps—1:12,500,000 or 197.3 miles to 1 inch for Canada as a whole; 1:6,336,000 or 100 miles to 1 inch for maps of eastern and western Canada and 1:2,217,600 or 35 miles to 1 inch for regional maps. The bases for the regional maps were derived from a map of Canada produced on the same scale in eight sheets on an oblique secant cylindrical projection with 110°W as the central meridian. There is one curious anomaly in the atlas. The limits of forest trees are shown on a map covering the whole of Canada on the 1:6,366,000 scale which meant that the map had to be folded twice to fit into the atlas cover! All but one of the city maps is on a scale of two inches to one mile (1:31,680). A total edition of 6000 copies were produced, 2000 being bound as a preliminary edition. Within two years it was listed as being out of print.

A second edition of the atlas appeared in 1915.[5] It was claimed to be a revised and enlarged edition, and certainly the information depicted was brought up-to-date. In 1912 the boundaries of Manitoba, Ontario and Quebec had been extended northward and these were incorporated into the maps. The results of the 1911 census were also used for the population maps. But in essence the atlas is very much like its predecessor with regard to scales and layouts (except that the limits of forest trees are this time shown on a 1:12,500,000 scale map!). In some instances the colours are changed to give the maps a more pleasing appearance, but the topics covered remain almost the same. A map of the world on Mercator's projection is added, but two maps on telephones which are in the 1906 edition are dropped. Consequently the 1915 edition had one less actual plate (excluding city plans) than the 1906 edition—34 as opposed to 35. The greatest change was in the coverage devoted to cities—the number of plates doubled from three to six partly to accommodate Edmonton, Regina and Victoria (the three provincial capitals which are not included in the 1906 edition) and partly to allow for the increase in the areal extent of the cities which had taken place in the intervening nine years.

Both atlases reflected a concern with communications. The 1906 edition devoted 14 of the 35 plates to telegraphs, telephones, railways, canals, lighthouses and sailing routes; the 1915 edition devoted 13 of the 34 plates to the same topics.

No new edition appeared following the 1921 census despite the slight renaissance in federal mapping which occurred in the 1920s (see Chapter 4). The 1915 edition was still in print and continued to be available in both bound form (for $3!) and as sets of loose sheets, at least up to 1927. The continuing federal concern and responsibility for the resources of Saskatchewan and Alberta is reflected in the atlas published in 1931 by the Dominion Bureau of Statistics entitled *Agriculture, Climate and Population of the Prairie Provinces of Canada—A Statistical Atlas Showing Past Developments and Present Conditions*. It included dot distribution maps based on statistical data from the 1926 census of the Prairie provinces and climatic maps produced by the federal meteorological service, and used base maps probably supplied by the Topographical Survey. The object of the atlas was 'to serve the purposes of the farmer, the prospective settler, the agricultural scientist, the business-man, and the student of economic geography as well as persons concerned with the settlement policy of the west'. Perhaps it was this statement of motives that was the greatest contribution of the 1931 atlas to the continued thinking about Canadian reference atlases. In 1934 Canada formally joined the International Geographical Union as a member state and in accordance with the IGU Statutes set up a National Committee. Part of this committee met in 1937 and suggested that a new atlas of Canada should be produced. But the project did not advance, presumably because national priorities were directed towards the economic depression of the 1930s and then towards World War II. However, in 1943, despite the fact that the war was still in progress, the Executive Committee of the Canadian Social Science Research Council appointed a three-man committee 'to explore the possibilities of the preparation and publication of a comprehensive Atlas of Canada'. Their report was published in late 1945.[6] It supported the project and concluded that the *Atlas de France* which had been in course of publication since 1933 was 'the best that one can see and use as a model'. It recognized the cartographic skills available in government departments, particularly in Ottawa. It also maintained that any new atlas should be 'scientific', i.e. a complex, comprehensive reference atlas presented in accordance with up-to-date geographical principles and based on adequate research. It was clear that there was a need to bring together three things: the cartographic experience, the wealth of information available in the federal departments concerned essentially with physical geography, and the data stored in the departments concerned with statistical, economic and social affairs. In 1947 a Geographical Bureau (which later became the Geographical Branch) was established in the federal Department of Mines and Resources, one of its aims being to produce a new atlas, for which approval was given in 1948.

The Third Edition of the National Atlas[7]

The intent of the new atlas was to show the nature, extent and use of the physical resources of Canada and their effect on the economy and society of the country. It was also to indicate the degree to which resources had been used, the way in which they had been exploited, the growth of the economy and the extent of social, cultural and political development reached by Canada in 1951. When viewed in sequence the maps in the atlas were to characterize and give meaning to the internal development of Canada and, to some extent, her international relations.

The first requirement[8] was a fundamental base map which would show the whole of Canada on one sheet, and a map drawn according to the Lambert conformal conic projection with 49° and 77° as the standard parallels, with a polyconic projection superimposed above 80°N latitude, was judged most suitable for the purpose. Printed at a scale of 1:10,000,000 this fitted onto a sheet 20 inches by 27 inches. Other maps in the atlas were drawn at scales which are simple multiples of 1:10,000,000—double (1:5,000,000); four times (1:2,500,000); ten times (1:1,000,000); half (1:20,000,000) or one-fifth (1:50,000,000). These scales were generally adhered to unless the topic warranted special consideration. For example, some weather maps required areal coverage of parts of the USA and Greenland as well as all of Canada and to achieve this a scale of 1:30,000,000 was required. Generally the 1:10,000,000 scale was used for a full-page map; the 1:20,000,000 where there were four maps to a page; and the 1:50,000,000 for a block of 20 small maps to a page. The 1:5,000,000 scale was used to show Canada in three parts—eastern, western and northern. The 1:2,500,000 scale was used for a detailed four-part map covering southern Canada, each map filling an entire page of the atlas—the Gulf of St Lawrence area, the Great Lakes area, the Prairies and the Far West. The most detailed sheet, one showing the physiography of southern Ontario, was on a scale of 1:1,000,000.

As an aid in clarity it was decided to print the bases in a 'faded' purplish grey colour rather than black which meant that thicker linework than that ordinarily employed was necessary. It also meant that when bases were required for which the faded colour was not suitable—e.g., the bathyorographic maps on which all water features and coastlines were printed in dark blue—it was necessary to redraft using finer linework and to use a different type face for the lettering. Three other basic colours were used: black, mainly for lettering; light blue for water-covered areas; and light buff for the remaining areas. Eight other colours (yellow, light brown, dark brown, green, light blue, dark blue, pink and red) were used in various combinations which, together with the use of four different screens (100 dots to the inch, 100 lines to the inch, 65 lines to the inch, and 30 lines to the inch) produced up to 30 tints. A few special maps, such as that showing bedrock geology and soil regions, required more than 30 tints, and special combinations of rulings and colours were used to convey the complex information. Most maps, however, required fewer than the full range of colours available.

It was also decided that the atlas should be in a looseleaf format, each sheet being bound by a linen strip so that it would lie flat when the atlas was opened and making it possible to remove the individual sheets for use if required. This and other physical aspects of the final work were very similar to the *Atlas de France* so highly commended as a prototype by the Committee of the Canadian Social Science Research Council.

The 1957 atlas was the first *Atlas of Canada* to be produced in French. Before this could be done three major problems had to be solved. The first was the problem of place names and the names of physical features. The guiding principle in establishing geographical names in Canada is local usage, and determining precedents for French usage was a time-consuming task entrusted to a special committee set up for the purpose. The second problem involved French equivalents for geographical terms used in the map legends and in the notes on the backs of the sheets. Because the scientific vocabulary of French-speaking Canadians is not always the same as that used elsewhere in the world—e.g., Canadians speak of the 'Région métropolitaine de Montréal' whereas in France the equivalent is 'Grand Montréal' or 'Montréal et banlieue'—it was necessary to consider each term or phrase individually. Finally, there was the problem of converting, in the most economic way, the material used to produce the English edition into reproduction material for the French edition.

The contents of both the English and French editions of the atlas were arranged so as to present Canada's historical, physical, human, economic, social and political geography in sequence. The first three sheets portray the origins of the country, one showing the routes taken by the principal explorers, the other two showing portions of the maps that were the products of the explorations. The lettering on these two sheets was done by hand imitating the style of the original maps, and they were hand-coloured as guides to the printers in the attempt to make the printed maps as much like the old maps as possible. The atlas then gives examples of current topographical sheets and aeronautical and hydrographic charts. These are followed by maps dealing with the physical aspects of Canadian geography—relief, geology, magnetism, tides—and 14 sheets on various aspects of climate. Maps of drainage basins, profiles of major rivers, soils and forestry follow and then, in turn, there are six sheets made up of small maps showing the ranges of representative insects, plants, trees, mammals, birds and inland fish. The section of the atlas dealing with human resources contains such subjects as distribution of the population, origins of Canada's people, main religions, and birth, marriage and death rates. The last section of the atlas shows the use of physical resources and the way in which cities, towns, rural municipalities and institutions are distributed as a result of this use. The economic portion of the section features maps of fisheries, sawmills, pulp and paper mills, agriculture, (including the distribution of farm animals and crops), mining, hydroelectric and thermoelectric power and manufacturing, ending with maps of transportation (canals, railways, airlines, shipping) and communication (radio and television networks) developed as a result of resource use. Finally, there are two sheets showing Canada's political evolution from colonial days to statehood and its overseas links through membership in the Commonwealth and the United Nations Organization and its participation in the Colombo Plan and North Atlantic Treaty Organization. The first of the 110 sheets was produced in September 1955 and the final sheet on 31 October 1958.

The Fourth Edition of the National Atlas[9]

Almost immediately plans were made for a fourth edition of the national atlas and eventually it was decided that it would be issued in four folios which could be stored in a specially designed, suitably titled box. This had a number of advantages. It obviated the delay that would be caused by producing a bound edition. Also a complete stock of loose sheets could be produced from the same press run. (This is not possible in a bound volume because, if the bound volume is to be portable, the material must be printed on both sides of the paper, which means that half of the maps will be split by the gutter between the separate sheets of paper and will be considered useless for distribution as loose sheets.) The advantage of an attractive bound volume would be offset by the lower price of the boxed version. Also individual maps could be removed for display or for study purposes and it would be possible to replace lost or mutilated sheets.

A page dimension of 14½ inches by 20 inches was chosen because it would be easily portable and would accommodate standard scales of 1:15,000,000 (for a two-page spread) and 1:7,500,000 (three sections for the whole country). As the latter scale had already been chosen for use as the basic scale for the *Atlas of the United States* and the page size was of the same dimensions as those used in the Surveys and Mapping's gazetteer atlas, it was felt that these choices produced a maximum amount of compatibility for users in, generally speaking, the same market.

The first folio was issued in August 1970; the second in May 1972 and the balance of the maps in February 1974. But by this time it had been decided to produce a bound volume of the separate sheets and this fourth edition of the *Atlas of Canada* was released to the public on 1 November 1974. Its joint release by Information Canada and the commercial firm, Macmillan of Canada, set a precedent as it was the first joint publication of such complexity between the federal government distribution agency and a commercial firm. The atlas was issued simultaneously in both English and French editions. As with the 1957 edition a bilingual format had been considered but rejected because it was felt the two languages used together would overcrowd the pages already filled with maps, legends, insets, charts and tables.

Of the press run of 16,000 copies, 13,000 in English and 3000 in French, Information Canada bought 5000, the Department of Energy, Mines and Resources bought another 1000 copies and the Macmillan Company retained the balance for sale through its normal outlets at $56 a copy. In order to attain wide distribution it was felt that the atlas should be of a size and weight 'to facilitate marketing and to convenience the reader in his home and the student at his desk'. Consequently it was decided to make the new atlas a more manageable size than its 1957 predecessor, which had cover dimensions of 16.5 inches by 21 inches and a weight of 17.5 pounds (42 centimetres by 53 centimetres, and 8 kilograms). Through extensive use of inserts, complex page layouts, sectioning of the country at scales larger than page-size scale, and a variety of cartographical refinements, the 1974 atlas, with a page-size scale of 1:15,000,000 as against 1:10,000,000 for its predecessor, sacrifices little of the detail and legibility of the 1957 atlas. The new atlas has a cover 10.25 inches by 14.75 inches which with a much more easily handled weight of approximately 5 pounds permits it to be carried in a briefcase.

The organization of the contents resembles that of the 1957 edition but nearly all of the data are new. Like the 1957 edition the material is divided into three main subject areas—physical, human and economic geography.

The physical geography section comprises 70 pages of maps, diagrams and explanations dealing with the physiography, geology, hydrology, soils, vegetation and climate of Canada. The 52-page human geography portion deals with the early discoveries and subsequent development of the nation, i.e. territorial evolution, population characteristics and changes, variations in education, religion and languages, and maps and charts regarding the aboriginal Indian and Innuit populations. Section 3, economic geography, is by far the largest. Its 132 pages deal with all aspects of the Canadian economy ranging from primary resources through aspects of production and communications to details of the retail trade. However, unlike its predecessor there is no gazetteer; nor are there any maps solely devoted to the location of cities, towns and villages. It was considered that this need was filled by the *Atlas and Gazetteer of Canada* and the *Atlas et toponymie du Canada*, with the same page size, issued by the same federal department in 1969.

The Fifth Edition of the National Atlas

It appears that the fifth edition will be unlike any of its predessors as it will consist of a number of folded sheets each within a wrap around cover of its own. Maps of the whole of Canada will be on a scale of 1:7,500,000, but there may be other maps, diagrams or explanatory notes on the same sheet. The only one to appear so far (1978) is 'Census Divisions and Subdivisions 1971'.

The fourth edition of the national atlas is obtainable from the Canada Map Office, Ottawa or The Macmillan Company of Canada. All previous editions are out-of-print although most large libraries have copies; the first (1906) edition is likely to be less frequently available.

Single Theme Atlases

The national atlases of Canada are characterized by the fact that their contents embrace many themes or topics tied together by related base maps and scales and the interrelatedness of the subject matter. There are some atlases which cover all of Canada, but are devoted to one main theme or topic usually applied or oriented to a particular group of problems.

In 1953 the Division of Building Research of the National Research Council and the Meteorological Division of the federal Department of Transport jointly published a *Climatological Atlas of Canada*. In fact it deals only with those aspects of climate which are related to building problems in Canada. There are 76 maps all consisting of isopleths printed on a common base map of Canada on a scale of 425 miles to 1 inch. The maps are grouped into eight sections: temperature, humidity, wind, snow, rain, sunshine and insolation, seismological disturbances and permafrost; in addition there is text, and there are tables and hythergraphs; altogether the volume extends to 253 pages. In 1957 this was followed by *Maps of Upper Winds over Canada* which was similar in form, scale and style. In 1970 the Meteorological Division published an *Atlas of Climatic Maps*. The maps cover the whole of Canada on a modified equal area projection and are printed in black and white. Full-page maps are on a scale of 200 miles to 1 inch (1;12,672,000). The following year the same service published *The Climate of Quebec—Climatic Atlas* consisting of 44 sheets of black-and-white maps and diagrams usually on scales of approximately 1:6,000,000 for the whole of Quebec and 1:3,000,000 for the southern part of the province.

In 1976 the federal Department of Agriculture published an *Agroclimatic Atlas* containing 17 maps showing the distribution of derived data of interest to agriculture (that is of indices obtained by interpreting the effects of weather and climate on plants and soils) for that part of Canada south of 60°N. The distributions are shown by means of isopleths with different colours between them. The maps are drawn on the Lambert conformal conic projection with 77°N and 49°N as the standard parallels on a scale of 1:5,000,000. At such a scale the individual map sheets are large. Consequently each is folded twice so that the volume is a more convenient size.

One of the largest single theme atlases is the *Inventory of Canadian Freshwater Lakes*, a boxed set of nearly 600 sheets issued by the federal Department of the Environment in 1973. It includes index maps for each of the six regions covering Canada showing the location and name of all freshwater lakes with a surface area in excess of 100 square kilometres. Each of the 562 lakes is shown on a separate sheet with its islands and other physical data.

Some of this information was used in the preparation of a *Hydrological Atlas of Canada* which was published by the federal Department of Fisheries and Environment in 1978 as a source of information for those involved in large-scale water planning and management and as a general educational and bibliographic reference document on water resources in Canada. The general style is very much like that of the third (1958) edition of the national atlas. The base maps are printed in grey and on scales of 1:10,000,000 for one map of Canada on one sheet and 1:20,000,000 when there are four maps of Canada to one sheet. The sequence of presentation of the 33 sheets generally follows the movement of water through the hydrologic cycle. The first is an artist's representation of the Hydrologic Cycle. The next 20 are maps of precipitation and its various aspects, updated versions of previously published information and with isoline values in metric units. The last 11 maps (except permafrost) are relatively new and cover such topics as water balance, suspended sediment concentrations, water quality of surface waters, surficial hydrogeology and bedrock hydrogeology. Each map is accompanied by a supporting text explaining the hydrological parameters depicted, the data used and its sources, and data not suitable for presentation on the maps. Related to this atlas is *Glacier Atlas of Canada* which is available as a set of unbound sheets of uniform dimensions—15¼ inches by 11 inches. It is actually a glacier inventory. Each sheet includes a map of an area in which perennial ice and snow is found. Each map uses deep blue for water areas and rivers and streams (usually left unnamed). Perennial ice and snow areas are indicated in light blue and each glacier is numbered in black. A small 'index [sic]' or location map is included on each sheet and the surround of the maps and the legend is printed in buff. The main maps are all on a scale of 1:500,000, but in order to fit the map areas onto the standard sheet the orientation varies; north is not always at the top of the sheet. In such instances the marginal latitude and longitude figures are often at an awkward angle to read. There are index maps for the larger areas which show the sheet lines for the detailed maps. For example, Plate 4–0, Axel Heiberg Island shows the sheet lines for the four 1:500,000 maps on which the individual glaciers are numbered. These index maps are on varying scales but 1:666,667 is generally suitable. The Baffin Island index however is on a scale of 1:4,500,000; that of southern Alberta on a scale of 1:2,000,000.

Single Theme Regional Atlases

There are a number of atlases issued by the federal government which deal with a single theme for part of Canada as opposed to the whole of the country.

One of the earliest of these was an *Ice Atlas of Arctic Canada* which was drawn and reproduced by the federal Department of Mines and Technical Surveys and published by the Defence Research Board in 1960. It consists of 29 maps of the north of Canada, essentially all the Arctic islands, on a scale of 80 miles to 1 inch (approximately 1:5,000,000). Each map summarizes for a particular date or range of dates every available sea-ice observation from 1900 onwards for 324 reference stations spaced 55 miles apart throughout the area. Sector diagrams are used to display what is known about ice conditions at a station; the size of the sector shows the proportion of observations of ice of a particular severity to the total number of observations available.

A *Sea Ice Atlas of Arctic Canada 1969–1974* was published by the Department of Energy, Mines and Resources in 1975. It depicts the results of aerial sea-ice observations in the general region of the Queen Elizabeth Islands and Parry Channel. The coasts are shown in outline on scales which vary with the area covered from approximately 80 miles to 1 inch (about 1:5,000,000) to 50 miles to 1 inch (about 1:3,000,000). The geographical distribution and extent of various types of sea ice and their characteristic features at different specific and identified times throughout the years are shown by means of colour patches and superimposed letter and figure codes.

Atlas of Eastern Canadian Seabirds was published by Environment Canada in 1975. It summarizes basic information on the ecology and breeding distributions for the seabirds of the Gulf of St Lawrence, the Atlantic provinces and the eastern Canadian Arctic. There are some 130 maps, all in black and white and all without scale although latitudes and longitudes are indicated.

A regional atlas dealing with several but related topics was published by the federal government in 1973 entitled *Gulf of St Lawrence Water Uses and Related Activities*. It consists of 10 sheets of maps in full colour each usually displaying some aspect of the whole region on a scale of 1:2,000,000. When there is more than one map on a sheet the scale of 1:5,000,000 is used.

Provincial and Territorial Atlases[10]

It was logical to move from the atlases of counties and their political subdivisions described in Chapter 2 to atlases of provinces with their major political subdivisions—the counties. It fell to Miles and Company, who had published the county atlases of Oxford (1876), Halton (1877), Peel (1877), Wellington (1877) and York (1878), to publish *The New Topographical Atlas of the Province of Ontario, Canada* in 1879. It is not a 'topographical atlas' in the modern sense of the word. There are 45 maps of the counties and five of the districts of Ontario. They show township boundaries, division court boundaries and numbers, churches, schools, post offices, railways, roads, and rivers and streams. The atlas also includes 49 town and city maps and the four sheets which cover Ontario from the 'New Railway and Postal Map of the Dominion of Canada' published in 1878 on a scale of 10 miles to 1 inch.

A similar publication in the same tradition was published in the same year (1879) by Frederick Roe of St John, New Brunswick. The *Atlas of the Maritime Provinces of the Dominion of Canada* includes maps of each county of each province (and the Magdalin [sic] islands) on a scale of seven miles to one inch (1:453,520) and 15 city and town plans, again on a uniform scale of 1200 feet to an inch. The maps of St John and Halifax show water soundings in figures. Each county map shows railways, roads, populated places, and named rivers and streams. Notable highlands are frequently (but not always) shown by hachures and named. However this atlas attempts to go beyond a mere collection of what were basically county plans by including a hand-coloured geological map of the three provinces on a scale of 1:584,000, with an inset of the island of Newfoundland on a scale of 50 miles to 1 inch. Part of the material in this atlas formed the basis of the *Atlas of Prince Edward Island* published in 1880 by J. M. Meacham and Co. of Philadelphia.

No further progress was made in the production of provincial atlases for more than 60 years. Ontario produced a publication which remotely resembled a provincial atlas in 1941, but the real development of provincial atlases did not come until after World War II when the knowledge that the federal government was producing the third edition of the national atlas, coupled with the stimulus of a Commission on National and Regional Atlases in the International Geographical Union, undoubtedly set the provincial governments thinking along similar lines.

Ontario

Ontario was the first province to issue an atlas published by a government department. In 1941 it produced the *Ontario Forest Atlas—For Schools*. It was an extremely modest effort of 14 sheets almost all devoted to forest composition, distribution and exploitation. Two scales were used: 145 miles to 1 inch and 125 miles to 1 inch. A second edition appeared in 1945. Since then two more editions have appeared with a change of title to *Ontario Resources Atlas* and sometimes with the addition of new subject maps. The 1958 edition contained 36 maps, the 1963 edition 33. The two scales used in the later editions were approximately 125 miles to 1 inch (for all or most of the province) and 55 miles to 1 inch for the southern part of the province. The base map shows some of the major rivers and has a graticule, and on it the thematic information is printed in colour.

The atlas is not organized according to the principles of geographic methodology, being concerned with the resources of Ontario which are of particular importance to the objectives of the Department of Lands and Forests. However the success of running to four editions is a measure of its usefulness during a period when little else of this nature was available.

There is no direct evidence to suggest that the *Ontario Resources Atlas* led to the *Economic Atlas of Ontario*, a much more ambitious work in the genre of the national atlases. It was published in 1969 by the University of Toronto Press, for the government of Ontario, the cartography and the research having been carried out by university personnel. Unlike several provincial atlases which preceded it and those which were to follow it, the *Economic Atlas of Ontario* departs from the traditional presentation of thematic data in both order and cartographic technique. Instead it provides analytical and synthetic insights into the fundamental geographical relations of the region so as to bring out the meaning and significance of the province and its parts. Modern computer techniques were effectively employed in manipulating data. The nine major sections into which the atlas is divided are: aggregate economy, population, manufacturing, resource industries, wholesale and consumer trade, agriculture, recreation, transportation and communication, administration, and reference maps. These cover essentially all of the important aspects of the economic life of the province, but not necessarily all the geographical essentials of the province. For example, there is not one adequate relief map in the entire atlas.

If there is one area where the provincial atlases differ markedly from the national atlas it is in the matter of scales. Seldom are scales selected so that they bear a simple numerical relationship with those of the national atlas or sister provinces. But Ontario appears to be almost completely oblivious to scale. The Ontario atlas does not use the representative fraction or the numerical method of expressing scale. When it shows any scale at all, it is linear (graphic) and is expressed in miles to the inch, the most common being 20 miles to 1 inch. The absence of metric units is particularly surprising as all the legends, titles and notes are given in both English and French. Also rare is the use of any graticule, or even border indications of latitude or longitude, or any sign that the north of the map is at the top of the page. To some, therefore, the *Economic Atlas of Ontario* is a series of cartograms. They are, however, cartograms of great use to the research worker for whom apparently they were mainly intended. The foreword to the atlas states that 'It should appeal particularly to students, teachers, government officials, business men and others engaged in research.' Almost as an afterthought it adds 'but there are contained in it maps and information which will be of interest to all residents of Ontario'.

The provincial government has also published (1976) the *Parry Sound District Atlas*, the first comprehensive atlas of an Ontario county or district since the appearance of the nineteenth-century county atlases to which reference has already been made. It contains 28 maps on a scale of four miles to one inch (1:253,440). The base map is printed in brown for roads, political subdivisions and concession numbers, and blue for water. The distribution of various phenomena is indicated by bolder colours. They cover the following topics: population distribution, landscape characteristics, resource capability, land tenure and subdivision, property assessment, economic activities, transportation and communication systems, and community facilities and services.

One other regional atlas of Ontario deserves to be mentioned, the *Ontario Arctic Watershed*. It was published by the federal government (Environment Canada) in 1975 and despite its title covers all of northern Ontario. It includes 23 maps in full colour most of which show information overprinted on a grey base on a scale of 1:10,880,000.

British Columbia

British Columbia was the first province to produce a modern reference atlas. Entitled the *British Columbia Atlas of Resources* it was actually published before the 1957 *Atlas of Canada*. But some of its editors had worked in Ottawa on the national atlas and many of the approaches to the British Columbia maps show similarities to the later atlas. The provincial atlas was conceived in 1954 at the Eighth Conference of British Columbia Resources, a conclave of technical and scientific personnel from government, university and industry, devoted to the analysis, conservation and rational exploitation of provincial resources. The atlas is, in effect, a cartographic summary of the findings of the Conference. Draughting, lithography, and printing were done by a commercial firm, with an editorial team composed of university and provincial government geographers.

A page of text accompanies each map page to explain and amplify the patterns revealed in the maps. This is the first time that a mixture of text, photographs and maps appears within an indigenous Canadian atlas. It consists of 48 plates embracing 91 maps and covers all of the principal topics which complex reference atlases usually include with the exception that it has no city maps. The largest standard scale is 1:3,484,440 (55 miles to 1 inch); other maps of the whole province are on a scale of 1:7,603,200 (120 miles to 1 inch). All the maps are on a polyconic projection.

A second edition simply entitled *Atlas of British Columbia* appeared in 1979. It is basically an up-date of the earlier atlas but the basic standard scale has been reduced to 1:6,000,000.

Manitoba

The *Economic Atlas of Manitoba* was published in 1960. Like the British Columbia atlas it was partly inspired by events in Ottawa as its editor had worked for the Geographical Bureau. Also like the British Columbia atlas it was sponsored by the provincial government and the major work was also done by university geographers. Lithography and printing were contracted to a commercial firm. In cartographic execution the Manitoba atlas was an advance over its British Columbia counterpart. It had fewer plates (37) but more maps (101) including four large-scale maps of metropolitan Winnipeg showing daytime population density, land use, urban growth and the distribution of 15 types of manufacturing plants based on field work carried out for the federal Geographical Branch, and as might be expected places greater relative emphasis on agriculture and climate. Fifty-three agricultural subjects using 37 maps and seven plates are dealt with under such headings as types of farming, the farm unit (e.g. number of farms and size of farms), crops, livestock, farm economic data (e.g. value of livestock per acre, value of farm buildings per farm, gross income per farm and non-farm labour), and farm tenure. Such treatment must be rated as specialized treatment, as must the climatic data which was primarily selected for its relevance to agriculture. The largest standard scale is 1:1,837,440 (29 miles to 1 inch) which is used for maps covering the southern half of the province. This section is also mapped on a scale of 1:2,957,920 (47 miles to 1 inch). Maps of the whole province use either 1:2,471,000 (39 miles to 1 inch) or 1:5,068,800 (80 miles to 1 inch). All the maps are on the Lambert conformal conic projection with 47° 30′ and 65° 30′ as the standard parallels, derived from the map of Manitoba prepared by the provincial surveys branch on a scale of 20 miles to 1 inch (1:1,267,200).

To Manitoba goes the distinction of having produced the first comprehensive reference atlas of any Canadian city, the *Atlas of Winnipeg*. Although assisted by the provincial government and compiled by geographers from the University of Manitoba it was actually published by the University of Toronto Press in 1978. The atlas consists of 67 plates of 151 maps. It is heavily oriented toward the social aspects of the population of the city and hence takes a cartogram approach, many of the sheets being essentially visual displays of census data. The scale of the maps of the city as a whole on one full page is approximately 1½ miles to 1 inch.

Saskatchewan

In 1969 there appeared two more provincial reference atlases produced in modern carto-geographical style: the *Atlas of Saskatchewan* and the *Atlas of Alberta*—remarkable achievements indeed for two provinces which did not even exist when the first *Atlas of Canada* was published!

The *Atlas of Saskatchewan*, published by the University of Saskatchewan, was far less dependent on the resources of the provincial government both in preparation and publication than any of the other atlases dealt with in this chapter. Its aim was simply stated: 'to portray the geography of the Province of Saskatchewan'. Its pedagogical bias is also evidenced by the inclusion of 139 pages of descriptive text, 44 coloured photographs of soil and vegetation types and several black-and-white illustrations. The maps in the Saskatchewan atlas are grouped into twelve sections: Introductory Physical Geography; Historical Geography; Population Geography; Physical Geography; Zoogeography; Forest Geography; Agricultural Geography; Minerals; Commercial Fish and Fur and Recreation; Industry, Services and Circulation; Urban Geography and Administrative Areas. The emphasis, however, as indicated by the length of each section is on physical geography, agriculture, industry, services and circulation. Most maps are single topic distributional displays which together are intended to provide an encyclopaedic approach to the overall geographical nature of the province. Maps of many of the topics are available in the national atlases or as separately published thematic maps, but if the *Atlas of Saskatchewan* is considered to have a textbook function this would amply justify the duplication of material appearing in the national atlas.

The most common scales used are 1:4,500,000; 1:6,000,000 and 1:9,000,000.

Alberta

The *Atlas of Alberta* like those of the other provinces was produced and published as a result of a cooperative partnership between university and government. Its aims too are similar: 'a reference for our own citizens in schools, homes and offices . . . and equally valuable to all those outside of Alberta who entertain an interest in this province'. Because of this it presents almost all of the mappable raw data for the province of Alberta following the order of geographical information which had by this time become traditional in such atlases: Relief and Geology, Climate, Water, Vegetation, Soils, Wildlife, History, Population, Land Use, Agriculture, Forestry, Fishing and Trapping, Minerals, Power, Manufacturing Services, Settlement Patterns and Administration. The most original and innovative section is that on Settlement Patterns. As might be expected in a province embracing some semi-arid areas, there are well-developed maps of water resources. The scale of the maps on a double page is 1:20,000,000 and on a single page 1:3,300,000. Unlike most of the provincial atlases but like the national atlases the *Atlas of Alberta* includes a minimum of text relying almost entirely on its map images to convey its message. It also uses a distinct cartographic style and its design and production standards are of high professional quality.

Northwest Territories

In 1947 the federal government published *Canada's New Northwest* a book which contains a series of seven maps devoted to the topography, geology, agriculture, forest types, water power and transportation of the area all on the same base map to a scale of 1:5,068,000. In their small way they could be considered the forerunners of the *Atlas of the Northwest Territories, Canada* published by the federal government in 1966. It covers the following categories: Political and Administrative Boundaries, Geology, Physiography, Climate,

Oceans, Vegetation, Wildlife and People. Interesting maps include three on exploration up to 1920, economic regions, missions and churches, Hudson's Bay Company establishments, native peoples, snow and ice cover. The scale of the maps is ca. 1:14,000,000.

Quebec

Quebec has published several sections of an economic atlas of the province. The first section *Atlas du Québec: l'agriculture* appeared in 1965. Most of the maps are on a scale of 1:4,600,000; others are on a scale of approximately 1:3,000,000. It consists of 47 loose sheets folded to fit into a wrap around cover. All the maps are in black and white. Three other sections in the same style have since appeared: 23 sheets on Industrial Activities, 32 sheets on the Tertiary Sector (Commerce, Construction, Finance, Services, Tourism, Transport and Communications) and 29 sheets on Population.

The government of Quebec has also been active in the publication of thematic and regional atlases. *Atlas Climatologique du Québec Température et Précipitation* appeared in 1978. It consists of 42 sheets of maps on which hypsometric tints indicate major differences in elevation with climatic isopleths overprinted in brown. The maps of central Quebec are generally on a scale of 1:3,000,000; sometimes they are on a scale of 1:6,000,000. The maps of northern and eastern Quebec are on a scale of 1:9,000,000. *Cartes de Pêches en 1976* issued by the Department of Industry and Commerce in 1977 consists of outline maps of the Gulf of St Lawrence or parts of it, drawn on a Mercator projection on which is overprinted computerized data from Quebec fishing trawlers related to fish types and fish stocks in the 1976 season. Somewhat similar is *Atlas Géochimique des sédiments de ruisseau. La Grand Rivière* and *Atlas Géochimique des sédiments de ruisseau. Région de Rouyn-Noranda* issued by the Ministry of Natural Resources in 1977. They consist of sections of the 1:125,000 topographical maps printed in a monochrome grey on which geochemical information is overprinted in solid colours—usually black and red.

The first of Quebec's regional atlases, *Centre du Quebéc Meridional*, appeared in 1963. It consists of 38 maps, generally in one colour on a scale of approximately 1:50,000. *Atlas Régional du Bas-St-Laurent, de la Gaspésie et des Îles-de-la-Madeleine* followed consisting of 67 sheets with maps on a scale of approximately 1:60,000 for the whole region on a single sheet to approximately 1:170,000 when several maps of the region appear on the same sheet. *Atlas Ezaim* (Atlas of the Ecology of the Zone of the Airport International of Montreal), financed by the federal government, was published in 1975 by Les Presses de l'Universite de Montreal. It consists of 55 maps, all in black and white on a scale of 1:50,000.

Newfoundland

In 1974 the government of Newfoundland published *Resource Atlas Island of Newfoundland* for which the cartography and reproduction were done by federal government agencies. It consists of a series of 14 maps on a scale of 1:1,000,000. They depict known capability for forestry, recreation, wildlife (ungulates) and agriculture, and resource use.

Nova Scotia

Between 1974 and 1975 the Nova Scotia Department of Development published five slim volumes of a *Nova Scotia Development Atlas.* The sections consist of 12 maps on forestry; 33 maps on agriculture; 22 maps on income and employment (in two volumes, each of 11 maps); 9 maps on secondary manufacturing and 36 maps on housing. The maps in each volume are in black and white, with one exception. When one map occupies a whole page it is on a scale of 28 miles to 1 inch (1:1,774,000); two to a page are on a scale of 46 miles to 1 inch (1:2,915,000); and three to a page, 66 miles to 1 inch (1:4,182,000).

In 1977 the Nova Scotia Department of Lands and Forests published *A Book of Maps*. It consists of 20 sheets each depicting some aspect of the resources of the province. Each is printed in several colours on a scale of 26.6 miles to 1 inch (1:1,685,270). None however have any grid reference or show latitude and longitude. The atlas also includes a cartogram showing, by means of flow lines of different thicknesses, the value of exports of forest products from Nova Scotia to various parts of the world.

15. Projections, Spheroids, Datums and Reference Systems

This chapter covers the following technical aspects of Canadian mapping: projections used in Canadian mapping; spheriods and datums; Canada's National Topographic System, and the UTM reference grid. There are numerous excellent texts available on the mathematical properties of projections and spheroids, so these subjects will be covered only in their special application in Canada. The National Topographic System is of course unique, so it will be explained in detail.

Projections

Projections have played a significant role in Canadian mapping ever since Champlain produced the first maps of this country that had any claim to accuracy. He started each map by first drawing a grid of latitude and longitude lines which then served as a rigid framework for his cartography. His 1613 map used a simple rectangular projection with both parallels and meridians being spaced to scale at the centre of the area of his explorations. Today such a projection would be called an equi-rectangular cylindrical projection. His map of 1632 shows considerably more sophistication. His lines of longitude converge toward the pole in proportion to diminishing distances between meridians which results in a sinusoidal projection, if modern terminology were to be used.

In the early days of land surveying and mapping in Canada, the whole concept of map projections was largely ignored for medium- and large-scale mapping. Township plans were drawn up on the presumption that meridians and parallels were straight lines at right-angles to one another. This convention provided an easy method of converting the survey notes into a simple map. The outlines of the township were fitted onto the paper in the most convenient layout and were annotated with magnetic bearings. Distances were scaled off in the chosen number of chains to the inch. The surveyors of course knew that meridians converge toward the north. In a hypothetical 10-mile 'square' township situated at latitude 45° with sidelines surveyed along meridians, the north boundary would be about 110 feet shorter than the south boundary. But the surveys of the day were so inaccurate that such a gentle taper of the township would never have been detected by the field surveys, and at the normal scale of 40 chains to the inch the difference in length would be smaller than the thickness of the line used to depict the township boundary.

The next step in mapping the settlements was to use the township plans as source material for district and county maps. These maps covered extensive areas on which the convergence of the meridians and the curve of the parallels could not be ignored. Information on the type of projection used for early county maps is very scarce. The statement in Roe's *Atlas of the Maritimes* (to which reference has been made in Chapter 2) that its maps were drawn on the rectangular polyconic projection was rare indeed. In almost all other mapping, in the days before this century, no projection information is given. However, as the maps generally have straight meridians and curved parallels (as they exist on the ground in the surveyors' environment), one is led to the presumption that a conic projection was used. One must again presume that a single standard parallel was used, as the benefits of two standard parallels were not required.

The earliest maps of the Geological Survey of Canada were of relatively small areas and local plane coordinates were used. Meridians and parallels, when shown, were straight lines at right-angles to one another. However, when larger areas, as in the Geological Survey one-inch series in Nova Scotia (published between 1877 and 1916), were being drawn a better projection was required. It is not known what projection was used for the first few sheets, but, shortly after 1884, when the United States Coast and Geodetic Survey published tables for a polyconic projection, this projection was adopted for maps of this series. The projection was based on Clarke's reference spheroid of 1866 and employed zones of 8° of longitude. The federal government was to continue to use this projection for virtually all large-scale mapping, both geological and topographic, until 1926. Canada's first native-born projection for medium-scale mapping was that used for the three-mile series of the Canadian Prairies. It is difficult to imagine any map projection ever designed that was more closely in tune with the surveyors working in the field. As has been mentioned in Chapter 3, the setting out of townships on the Prairies was governed by baselines and control meridians that were surveyed with special care and precision. The baselines were surveyed as straight-line chords of parallels of latitude, each chord being six miles in length (i.e. the width of one township). This means that the surveyors, as they ran the baselines westward across the Prairies, would deflect their line northward by a small angle every six miles. For example, the 18th baseline touches latitude 54°56′17″ every six miles. To maintain this contact the surveyor had to turn his line northward by an angle of 7 minutes 29.5 seconds at each six-mile point.

To the surveyor in the field, parallels of latitude were seen to be curved lines while control meridians, surveyed every 4° of longitude, were as straight for their whole length as it was possible to survey them. A simple conic projection was chosen for the maps of these surveys so that the straight meridians and curved parallels would be maintained. The method of developing the graticule for any particular sheet was simplicity itself, though it must be admitted it could not produce the precise results that are demanded today. The sheets, one should recall, were laid out so that each contained three baselines; the top and bottom baselines formed the north and south boundaries of the map (excluding the overlap with adjoining sheets) while the third baseline ran across the centre of the sheet.

Two sheets, each 2° in longitude, were fitted between control meridians, so that a control meridian formed either the east or west boundary of each sheet (again excluding the overlap). To draw his graticule the cartographer would first draw the sheet's central meridian. On this he would then mark out to scale (sheets were drawn at two miles to the inch, but reduced photographically for printing to three miles to the inch) the intersections of the three baselines. They would be 24 miles (12 inches) apart. A line at right angles to the central meridian was then drawn across the paper at the intersecting point of the sheet's central baseline.

Each draftsman engaged in this work was provided with the graph shown in Figure 36,[1] and a set of the *Smithsonian Geographical Tables.* The horizontal line would then be marked off in increments of six inches, which at the drawing scale of two miles to the inch would represent 12 miles, or two ranges. The number of the baseline was of course known, as they were numbered consecutively north from the US border. The abscissae could thus be read directly from the graph for the appropriate range-marks out in both directions from the central meridian. The curved central baseline was then drawn by joining the tops of the abscissae. As this was a simple conic projection, the rest of the graticule could be easily completed by marking off the central meridian and central parallel at true scale using figures for the distance of minutes of latitude and longitude taken from the *Smithsonian Geographical Tables.* (The *Dominion Land Surveyor's Handbook* has always carried a table giving the latitude of the baselines, and the central meridian of each sheet was always set at an even degree of longitude.) The remaining meridians were then drawn at right angles to the curved central baseline and marked off to scale to provide intersection points for the top and bottom baselines. Generally, the two sheets with the same central baseline running between two control meridians were laid down at the same time so that the townships could be scaled off every three inches (six miles) from the easterly control meridian. The map detail was then plotted from the 1:31,680 township plans.

In this projection, scale was correct along meridians and along the central baseline, but the scale errors along the outside baselines were very small and in most cases were less than the tolerances allowed for the drafting.

In 1903 the first of the Chief Geographer's 1:500,000 series was published. No information on the projection used seems to have survived, but on examination the projection

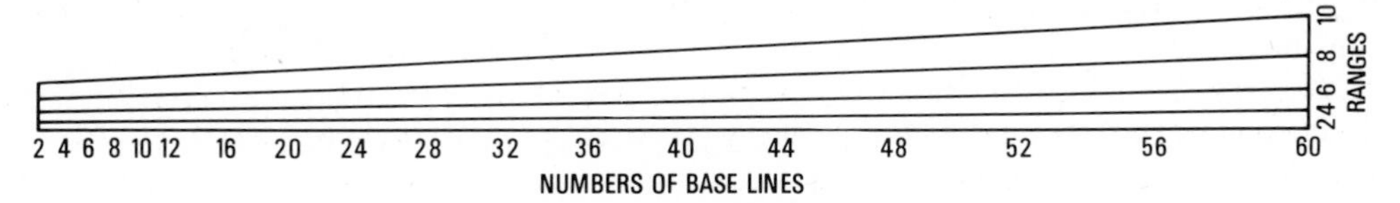

SCALE OF CO-ORDINATES

For curves of base lines on sectional maps

ORIGINAL SCALE: 2 MILES TO AN INCH (example is 1/3 of correct size)

HOW TO USE THE SCALE

On sectional maps the base lines are a series of chords, each chord crossing two ranges. These are plotted from a central meridian and a horizontal line at right angles to it.

The perpendicular distance (the co-ordinate) may be taken directly from this scale which is adapted to a map of two miles to an inch.

The graduations on the vertical lines are the co-ordinates for 2, 4, 6, 8, and 10 ranges on the several base lines numbered on the zero line of the scale.

Figure 36 Graph for plotting three-mile projection

appears to be the same as that used in the three-mile series. The meridians are straight, the parallels are curved, and an overlap with adjoining sheets allows a rectangular frame while still using certain meridians and parallels as the true boundaries of the sheet. As the drafting offices of the two series were in the same department, it seems probable that similar drawing systems were used for both series.

The three-mile projection was quite sufficient for scales of 1:190,080 and smaller, but it was unsuited for one-inch mapping. So, throughout the early years of the twentieth century, the American polyconic projection was used for large-scale mapping by the Military Survey, Geological Survey and the Topographical Survey. The polyconic projection was laid down in much the same manner as the three-mile projection. The draftsman would start by drawing the central meridian of the sheet and the parallel nearest the centre of the sheet. Using tables, and knowing the latitude of the central parallel, the coordinate values in inches of the required latitude and longitude, intersections were found and plotted. On this framework the map was built.

Unfortunately, the polyconic projection is not a conformal projection. As the surveys of the Canadian mapping agencies become more extensive, the requirement for a conformal projection become more important. Dr Deville, the Surveyor-General, commenced a search for the ideal projection for Canada in 1922, and shortly before his death in 1924 he decided on a Transverse Mercator projection with 8° zones. F. H. Peters, who succeeded Deville as Surveyor-General, continued this work. In 1924 tables were computed for plotting 8° TM sheet graticules, and in 1926 the first sheet on the projection was published (sheet 52 M-V, Kamloops, now 92 I/9).

The publication of sheet 52 M-V was more than a new topographic map on a new projection. It was the beginning of Canada's National Topographic System which will be described in detail later. The Canadian Army agreed to and supported the NTS (as the National Topographic System was called almost from the beginning), but they objected strongly to the Transverse Mercator projection. The main objection was the number and position of the zone boundaries which occurred (and still occur) at intervals of about 450 kilometres across Canada. There were 10 of these zone boundaries across the country and some happened to fall at points that were 'sensitive' to military operations, particularly those near Hamilton, Quebec and Halifax. To the military mind, Canada was a long thin country situated in an east–west direction, a state of affairs that demanded a conformal conic projection.

Lengthy pleas and expositions were made to Deville by officers of Canadian Army Headquarters, but he was firm in his decision to adopt a Transverse Mercator projection.[2] The Army alternative[3] was the Lambert conformal conic with four horizontal zones:

Zone 1 latitude 41° to 45°
Zone 2 latitude 45° to 49°
Zone 3 latitude 49° to 54°
Zone 4 latitude 54° to 59°

To reduce the scale error (i.e. the difference between the distance between two points measured on the map and between the same two points measured on the ground) a secant zone was employed, thus giving two parallels with zero scale error. These parallels are not used in working the projection, and neither were they used in the initial design of the system which was as follows: the central parallel of Zone 3 was reduced from a scale factor of 1.0 (i.e. true scale) to 0.995. This had the effect of reducing the scale factor of the north and south boundary parallels bringing them closer to unit. As it was essential in this system that map detail should fit across zone boundaries, the south parallel of Zone 4 was brought to the same scale as the north boundary of Zone 3. The north parallel of Zone 2 was likewise brought to the same scale as the southernmost parallel of Zone 3. Finally, Zone 1 was fitted to Zone 2 in like manner. The maximum scale error of the whole system was less than 1:2000 which made it suitable for all military purposes. Although the map detail was continuous across zone boundaries, the rectangular grid was not, and the computation of bearings and distances across zone boundaries required considerable calculation and the use of special tables.

To keep the divergence of grid north from true north within reasonable limits, three reference meridians were used:

Maritime (longitude 64°)—used for Zones 2, 3 and 4
Eastern (longitude 77°)—used for all Zones
Western (longitude 133°)—used for Zones 3 and 4 only

So ultimately the military proposal had nine grid discontinuities, and the part of Canada north of 59° was considered inconsequential for purposes of military operations. But the real disqualification of the military system, in civilian eyes, was the divergence of grid and true north. Even with three reference meridians this amounted to over 20 degrees in certain parts of Canada.

The conflict was not resolved, and in 1932 the military proceeded to 'go it alone' with the Lambert conformal projection. The grid numbering system adopted was called the Modified British Grid which consisted of grid lines 1000 yards apart numbered from 0 to 99 in both eastings and northings. Each 10,000-yard square bore an identifying letter. The grid reference of a point would thus consist of a letter followed by three digits of eastings and three digits of northings. The Modified British Grid could be applied to maps of the other agencies produced on the 8° TM with no apparent error, and a number of such sheets were gridded by overprinting during the Second World War. On examining many of these sheets today, one must agree that Deville was right; the oblique grid is a distracting feature, especially in a country where vast areas have been laid out with roads and lot-lines in the north–south and east–west directions.

The competition between the conformal conic and the 8° TM ended abruptly in 1949 when it was decided that all large- and medium-scale mapping done by the federal government would be done on the 6° UTM projection. Tables and instructions for this projection were provided by the US Army.

The Lambert conformal and both the 8° and 6° TM projections are projections that can be given a comprehensive rectangular grid over a complete zone. In all three projections mathematical tables were provided so that a point of known latitude and longitude could be converted to a grid reference.

This feature is used by the cartographer in drawing a map on the projection. The grid values of the geographic graticule and all points of the field survey that are to be used as a framework of the map are computed. In the early days the cartographer was provided with specially prepared paper that was glued down to a thin sheet of metal to give it dimensional stability. A very fine grid was then printed on the paper in a light non-photographic blue, and on this grid he would lay out the projection and control points of the sheet he was drawing. All other detail would be keyed to this control. Today an automatic drafting machine draws the projection and control points on the drafting medium which is usually a sheet of dimensionally stable plastic.

Today there are a few 1:50,000 sheets remaining in the NTS that have not been recompiled since the 8° TM days, but these are all now gridded with the UTM grid system. The 8° TM, however, lives on in the 1:500,000 series (see Chapter 8). A special conformal conic projection is used for the 1:1,000,000 air charts in which the two standard parallels are one-sixth of the latitude extent of the sheet inside the north and south edges of the sheet. (Each sheet covers 4° of latitude so the northern standard parallel is 40 minutes south of the north boundary and the southern is 40 minutes north of the south boundary.)

The Lambert conformal conic has been used extensively for general maps of Canada at small scales. The usual standard parallels are 49° and 77°. This projection gives a relatively undistorted picture of Canada, and as few people use such a map today for taking off precise areas, there have been few objections to its use. On most official maps showing the whole of Canada, the meridians north of 80° have been given an artificial inward curve so that they converge rapidly to the North Pole. This is done to allow the illustration of Canada's claim to sovereignty of the territory north to the Pole on the sector principle.

Canada's official *Map of the World* is drawn on Van der Grinten's projection. In this, the whole sphere is projected onto a circle. It is not orthogonal, nor is it equal-area. Probably it can be described as one of the least offensive projections available which shows the whole world without interruptions and without the excessive distortions of the Mercator in high latitudes. It has one peculiar property of no importance in this day of automatic co-ordinatograph plotting tables. It can be drawn using nothing more than a straight-edge and beam compass. Instructions for its construction are given in *An Introduction to the Study of Map Projections* by J. A. Steers, published by the University of London Press.[4]

In closing this section on projections, a short description of the use of projections by the provinces must be made. In every province some use has been made of the polyconic projection and in some this use continues to the present time. The tables for drawing this projection were easily obtained from the United States, and in provincial work before the Second World War there was no need for a conformal projection. But, this changed after the War when the provincial land survey associations began pressing for the adoption of conformal projections with a comparatively small-scale error. A scale error of no more than 1:10,000 was generally cited as a requirement, and this ruled out the UTM which because of its six-degree zones has a maximum scale error of about 1:2000.

In 1945, the Geodetic Survey prepared a plane coordinate system for Prince Edward Island based on the Lambert conformal conic projection with standard parallels at 46°09′53″ and 46°50′. The maximum scale error was about 1:30,000, but the projection was never used.

In 1959, the Geodetic Survey devised a plane coordinate system for New Brunswick based on the stereographic projection which is suited to that province because of its compact shape. In this projection, the projecting plane can be made secant to the spheroid, and in the New Brunswick experience the circle of true scale has a radius of 75 miles. The scale error does not exceed 1:10,000 for any point within 110 miles of the point of origin and at the extremes of the province it does not fall below 1:5000.

During the 1960s the federal government suggested to the provinces that a Transverse Mercator projection with 3° zones would serve both the needs of land surveyors and the producers of large-scale topographic maps. This projection was adopted enthusiastically by Quebec, Ontario and Alberta, and rather reluctantly by Nova Scotia, Manitoba and

Saskatchewan. Both the Yukon and Northwest Territories are far enough north so that the zones of the 6° TM are narrow enough to produce a scale error of less than 1:10,000. British Columbia continued with the polyconic projection for most thematic maps and the 6° UTM for their topographic work.

In recent years some of the provinces have had second thoughts about the correctness of using a different projection from that employed by the federal government. The 6° UTM can produce survey values well within the 1:10,000 limit if careful use is made of scale factors during survey computations. The great advantage in being on the federal system is that masses of federal data coming from Statistics Canada, the Geological Survey, Agriculture Canada, and other agencies is now being coded on UTM coordinates. There is some thought that when the changeover to the 1983 North American Datum (which will be described in the next section) is made a change to the 6° UTM should be made at the same time.

Spheroids and Datums

In geodetic surveying and topographic mapping the flattening of the earth at the poles must be taken into account when computing the results of long measurements made over the surface of the earth. The geodesists who made these calculations usually employ a spheroid approximating as closely as possible the size and shape of the earth. On such a mathematically definable smooth surface the bearings and distances of the surveyors can be transformed into positions of known latitude and longitude.

In Canada the Clarke reference spheroid of 1866 has been used since the publication of tables for this spheroid by the United States in 1884.

Colonel A. R. Clarke (1824–1914) was an officer in the Royal Engineers who served most of his military career in the Ordnance Survey. Although he excelled in many fields of surveying and geodesy, he is today best remembered for his three determinations of the dimensions of the earth. He made these in 1858, 1866 and 1880, with each successive determination being calculated from the results of additional survey data obtained from many parts of the world.

In the USA the surveys made before the Civil War were computed on the Bessel Spheroid of 1841. In the years immediately following the Civil War a number of long arcs of both meridians and parallels of latitude were measured, and on reducing these it was found that Bessel's dimensions were too small for a consistent fit. The Clarke 1866 Spheroid was in closest agreement to the American data and in 1880 it was chosen by the US Coast Survey as the American reference spheroid. As many of the Canadian surveys were tied into those made in the USA, and as Canada made extensive use of American textbooks, it was only logical for Canada to adopt the same spheroid.

The Clarke 1866 Spheroid has served the country well over the years, but with the advantage of satellite geodesy it is now recognized that an improved mathematical shape of the earth can be computed. Geodesists throughout the world are working on this project and expect to have an answer soon.

The vertical datum of Canadian maps is mean sea level. Canada maintains a number of tide gauges on all three oceans, and the surveys made for mapping are all tied to this vertical control net. As a matter of interest, it should be noted that the blue line indicating the coastline on Canadian tidewater sheets is the mean high water line, not mean sea level.

This means that the vertical interval between the shoreline and the first contour is always less than the other contour intervals.

The horizontal datum on Canadian maps is that provided by the 1927 North American datum. In 1927 geodesists from both the USA and Canada assembled all available geodetic data, which included a large number of positions the latitude and longitude of which had been obtained both by astronomy and by overland measurements. The whole system was then adjusted to make the discrepancies between the two types of position fixes as small as possible. When this was done a station near the centre of the continent and with a very small difference between its geodetic and astronomic position was chosen as the point of origin for all geodetic surveys of the whole continent. This station was at Meade's Ranch in Kansas.

In recent years the 1927 North American datum has been found wanting when used with modern survey equipment of high precision, and a new North American datum will be computed in 1983 which will be tied to the new spheroid which will be available at that time. The new point of origin will be the centre of gravity of the earth rather than a survey station on the earth's surface. The axes of 'earth centered coordinates' will be the polar axis of the earth and two lines at the earth's centre at right angles to the polar axis and to each other. One of the lines at right angles to the polar axis will pass through the Greenwich meridian. These scientific coordinates will of course be converted to equivalent values of latitude and longitude, or to UTM coordinates, for use in practical matters on the surface of the earth. There will, however, be small but visible shifts in the graticule and grid on Canadian maps when the 1983 North American datum is adopted.

Canada's National Topographic System

Canada's National Topographic System was designed to provide for the orderly mapping of the country at a range of scales from 16 miles to the inch to one mile to the inch. In the original concept each map at each scale would have its place in the overall grid, and thus important areas could be mapped first, at the scale required, and still be fitted into the system. A map numbering system was evolved which allows the informed map-user to distinguish from the map number the scale of the map and its geographical position. The map number also indicates what alternative map coverage exists at smaller scales and aids in searching for more detailed mapping at larger scales.

The NTS had a rather uncertain start in life. As has been described in Chapter 10, the Chief Geographer in the Department of the Interior was instructed to bring some order out of the chaotic mapping situation in the medium-and small-scale range. His solution was a map series at 1:500,000, each sheet covering 3° in both latitude and longitude. Each of these maps would then be quartered into 1:250,000 sheets designated NE, NW, SE and SW. There was no suggestion, then or later, to carry this quartering on into larger scales. For one reason, the Chief Geographer had no field staff to provide the topographic data for larger scaled mapping; for another reason, the 3° primary quadrangle does not lend itself to repeated quartering. In 1904 when the military started the first of Canada's large-scale topographical series, the Chief Geographer's system was ignored completely and a simple set of sheet lines for the military one-inch mapping was designed in which a 1° quadrangle was divided into eight sheets each measuring 15 minutes in latitude and 30 minutes in longitude. Within this simple lattice, the sheets were numbered more or less in the order in which they were produced.

	79	78	
46		etc.	
		4	
		3	
45	1	2	45
	79	78	

The three-mile sectional maps of the Prairies were units of a thriving medium-scale series quite apart from any other mapping in Canada. By 1905, when 64 sheets of this system had been published, a numbering system was designed (as described in Chapter 3) so that again the informed map-user would know by sheet number the sheet that was to the east or west, north or south, of the one he was using. But this system was based on the Prairie township system and each sheet extended over eight townships in the east–west direction and four townships north–south. This gave sheets the 'uncomfortable' dimensions of 42′30″ in latitude and 2° in longitude. This was obviously not suitable for a national system.

In 1922 the Geographic Board of Canada made a valiant effort to index all of the important maps available in Canada at that date. It produced a catalogue and a set of 10 index maps showing the location and extent of over 2000 maps. The index maps are so confusing that one researcher was moved to describe them as 'resembling the webs of a disturbed spider'. Figure 37 is a section taken from one of these maps. The catalogue and indexes were of course useful, but their greatest achievement was to force the Department of the Interior into action on a national topographic system. In 1925 the first version was launched. It was based on the *International Map of the World* numbering system (which is described below) which has an orderly array of primary quadrangles of 4° in latitude by 6° in longitude. This system was tried for two years, but in 1927 was modified to what is essentially the present system.

The reasons for change were threefold. First, a 4° by 6° format does not divide into quarters easily, and it had been decided by the Department of the Interior that a range of scales would be provided in the system, each half the next largest. Thus quartering would be required to produce scales of 1, 2, 4, 8 and 16 miles to the inch. Second, in a northern country the 6° longitude extent of the primary quadrangle tends to produce small maps. Finally, the mathematicians in the Department of the Interior had been searching for a suitable projection for this new topographic system and had finally decided on the Transverse Mercator. To keep the number of zones to a minimum and yet keep the scale error within bounds, it was decided to use zones of 8° of longitude. What better solution could there be than organizing the new system into 8° columns, and within the columns have primary quadrangles every 4° of latitude? Thus there would be no TM zone changes within any of the sheets of the system. The primary quadrangles are illustrated in Figure 44. The further division for sheets of larger scales will be explained later.

As has been mentioned in other chapters, the scales have been changed to metric equivalents, and the 8° TM has been abandoned in favour of the 6° UTM. But, other than minor alterations, the 1927 concept exists today. As these minor alterations have caused some unusual map numbers which still are held in many map libraries, they will be briefly described.

It will be noted in Figure 41 that the width of the mapped area of each sheet in the system

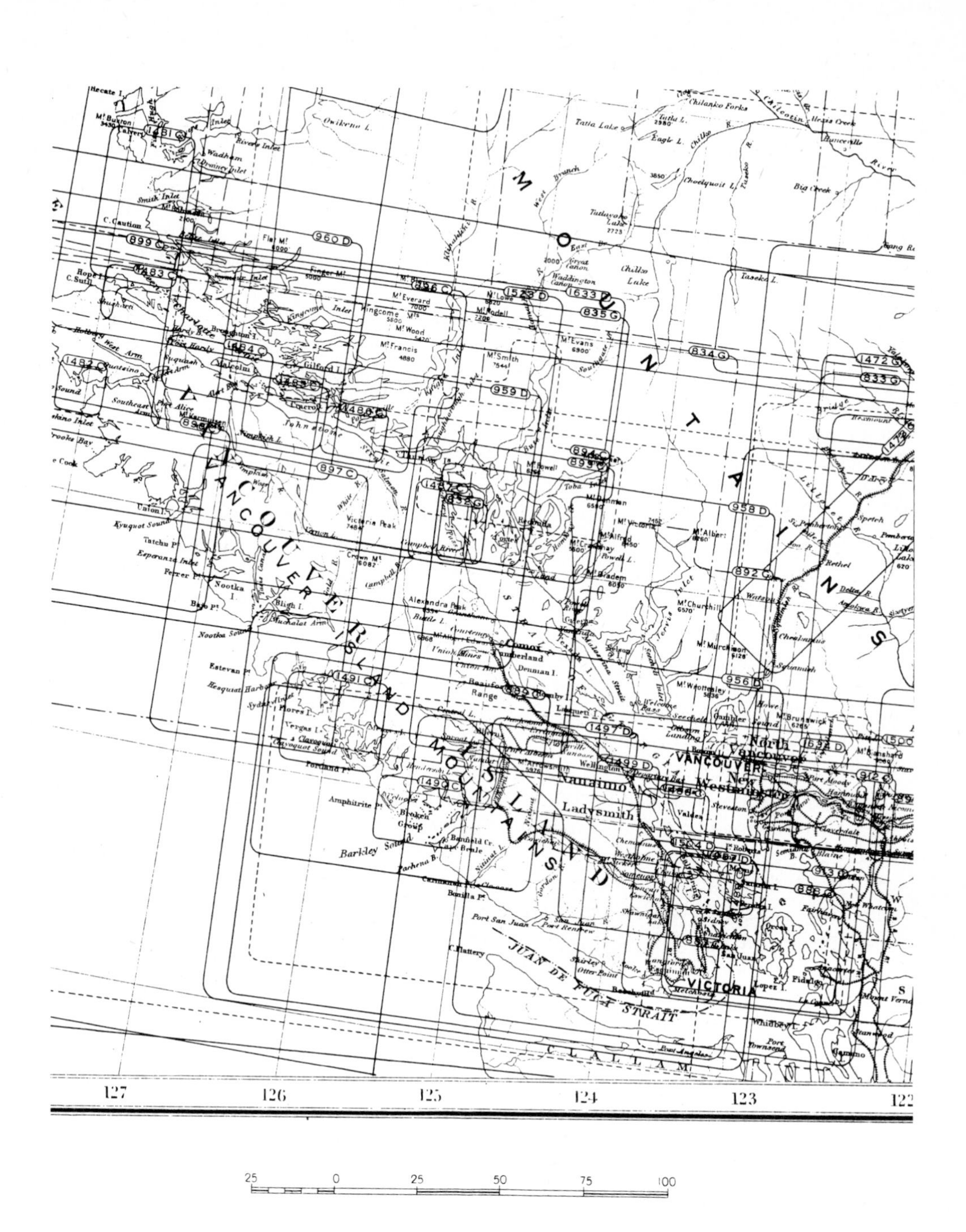

Figure 37 Portion of 1922 index

decreases northward with converging meridians. At 68° north latitude the sheets have become so narrow that they are doubled in longitude coverage. For example, the 1:50,000 sheet goes from 30 minutes to 1°; the 1:250,000 series goes from 2° to 4°. This doubling takes place again at 80° north latitude, but being this high in the Arctic relatively few sheets are affected. In the days of the publication of half-sheets in the 1:50,000 series, it was found that at 62° north, the longitudinal extent was such that the whole 30′ sheet could be squeezed onto the approved paper size. Then the Committee on Map Standardization had the idea of reducing the longitudinal extent of sheets between 60° and 64° north to 20 minutes, and placing 24 of them in each 1:250,000 quadrangle.[5] Needless to say, this created great confusion as many sheets in this band had already been published with the former NTS numbers and they now had to be reissued with new numbers. In 1960 it was agreed by all that this experiment was not working, and the system was returned to normal.

The present system of map scales and sheet lines used in Canada's NTS is quite methodical being based on a grid of primary quadrangles each 4° of latitude by 8° of longitude. Each quadrangle is a map of the 1:1,000,000 series and each is given a number as shown on Figure 43. For example, the Winnipeg Sheet is 62. The basic system for the larger scales is as follows.

1:500,000

Each primary quadrangle is quartered into maps of the 1:500,000 series and each of these is designated by the primary quadrangle number followed by NW, NE, SW or SE; e.g., 64 SW.

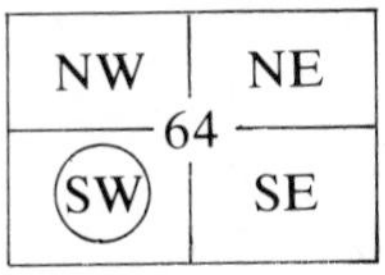

NW	NE
(SW)	SE

64

1:250,000

The primary quadrangle is also divided into the 16 sheets, lettered A to P, of the 1:250,000 series. Each sheet is designated by the primary quad number followed by a letter; e.g. 64D.

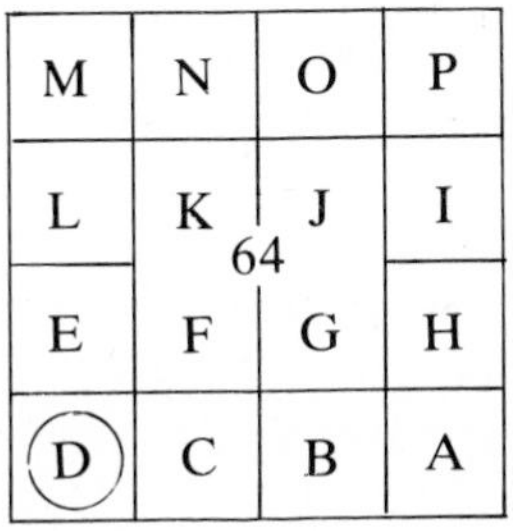

M	N	O	P
L	K	J	I
E	F	G	H
(D)	C	B	A

64

1:125,000

Each of the 1:250,000 quadrangles is quartered into maps of the 1:125,000 series and each is designated by the 1:250,000 quadrangle followed by NW, NE, SW or SE; e.g. 64D/SW.

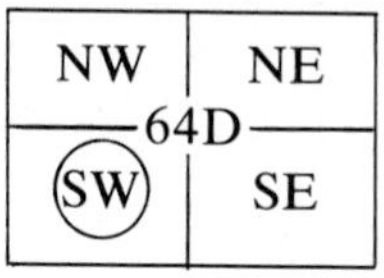

64D

NW	NE
(SW)	SE

1:50,000

The 1:250,000 quadrangles are also divided into the 16 maps of the 1:50,000 series numbered 1 to 16. Each sheet takes the 1:250,000 designation followed by a number; e.g., 64D 4. Many 1:50,000 sheets published before 1967 were divided into East and West halves. The serial number for such a sheet would have the letter E or W added; e.g. 64D/4E.

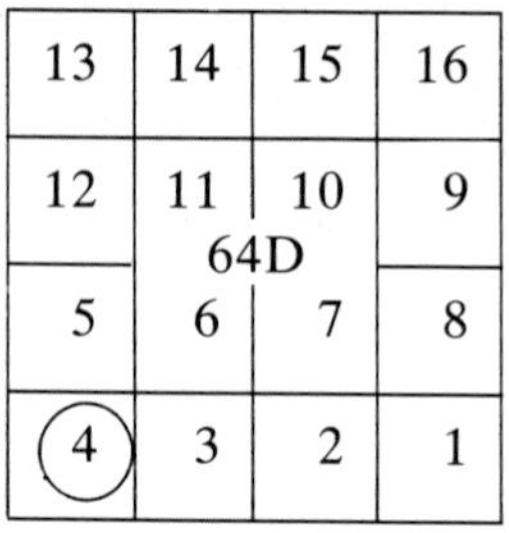

64D

13	14	15	16
12	11	10	9
5	6	7	8
(4)	3	2	1

1:25,000

In certain settled areas of the country 1:25,000 maps are published. In such cases the 1:50,000 quadrangle is divided into eight sheets letters a to h; e.g. 64D/4 d.

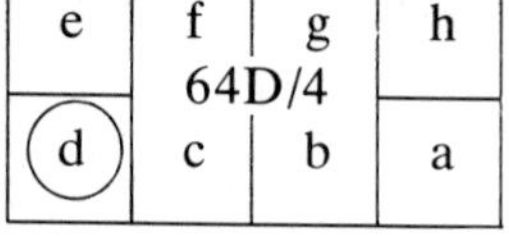

64D/4

e	f	g	h
(d)	c	b	a

As Canadian maps are bounded on their east and west sides by meridians it follows that as one goes north the convergence of the meridians toward the Pole causes a steady decrease in the area covered by a single sheet. At the point where the sheet becomes too narrow for practical purposes, the longitude span of a single sheet is doubled. Figure 38 illustrates a column of primary quadrangles in which the doubling (and concomitant relettering) can be observed.

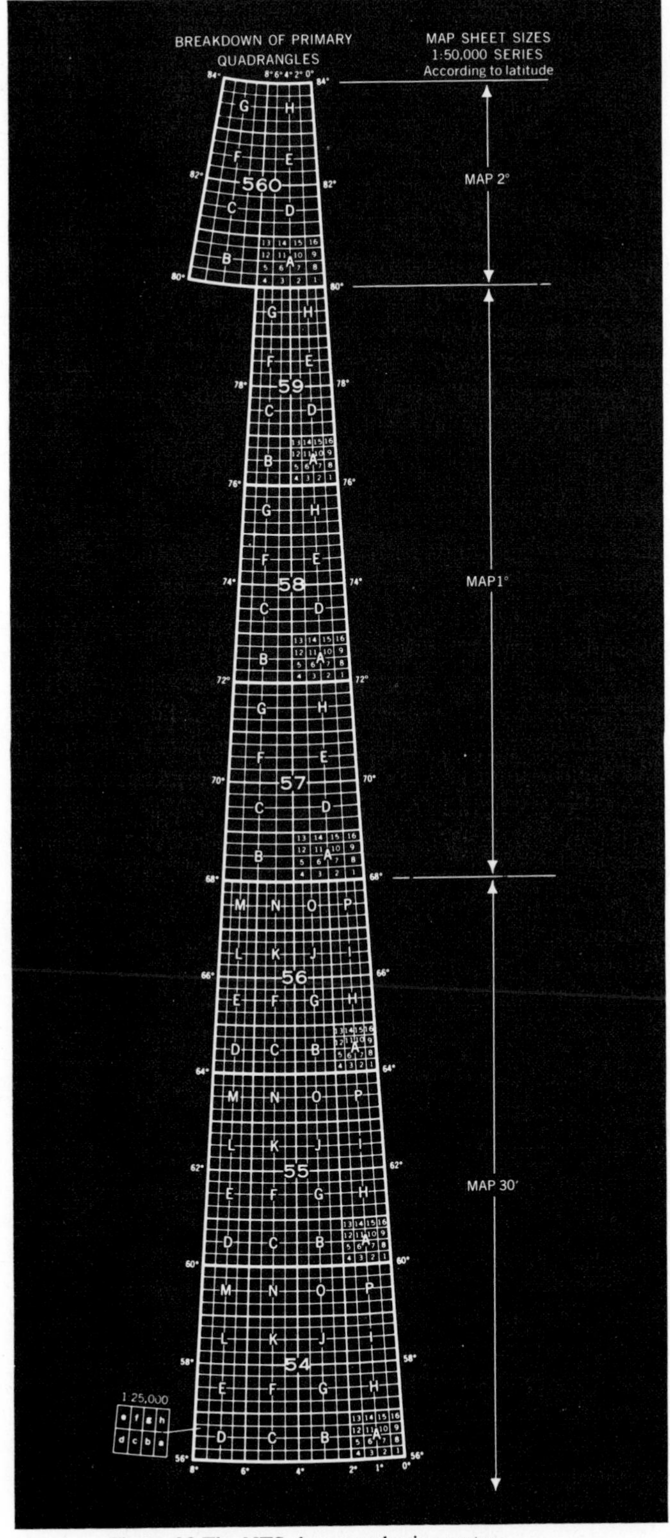

Figure 38 The NTS sheet numbering system

The International Map of the World System of Sheet Lines and Map Numbering

The IMW sheet line system is used by many countries (e.g. Australia, the Soviet Union, etc.) for defining and numbering topographic maps in the progression of scales from 1:1,000,000 to 1:25,000. In Canada the system is used for two series only, the IMW series itself (Chapter 10) and the military 1:250,000 JOG series (Chapter 8).

It has been noted that the basic dimensions of the IMW sheets are 4° of latitude by 6° of longitude. This provides the fundamental grid which has its origin at the intersection of the Equator and the 180th meridian. In the IMW alpha-numeric system all sheets lying to the north of the Equator start with the letter N. The next letter in a sheet designation indicates in which 4° latitude band the sheet lies. As there are 22 bands from the Equator to 88° north, the letters run from A to V. (The single sheet at the Pole is designated as NZ.) The two letters in a sheet designation are followed by a number, 1 to 60, which indicates the 6° column in which the sheet lies. The column numbering increases eastward from the 180th meridian. A typical designation of a Canadian sheet in the system is NN–11, Lesser Slave Lake. The Canadian layout of IMW sheets is shown in Figure 31 of Chapter 10, and it will be noted that north of 60° some liberties have been taken with the sheet lines to provide a more efficient east–west coverage. In dividing the 1:1,000,000 quadrangles into 1:250,000 sheets the IMW system calls for sheets lying between 40°N and 60°N to be 1° in latitude by 2° in longitude (i.e. the same dimensions as the NTS 1:250,000 sheets). As the basic grid is two degrees narrower than that of the NTS, the division (and numbering) is done as shown below, for example, NN–11–10

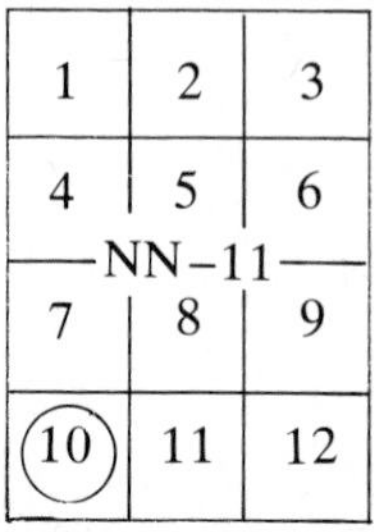

1	2	3
4	5	6
	NN–11	
7	8	9
(10)	11	12

Between 60°N and 72°N a 16-sheet division is made to the primary quadrangle, with the sheets being numbered, for example, NP–11/12–13

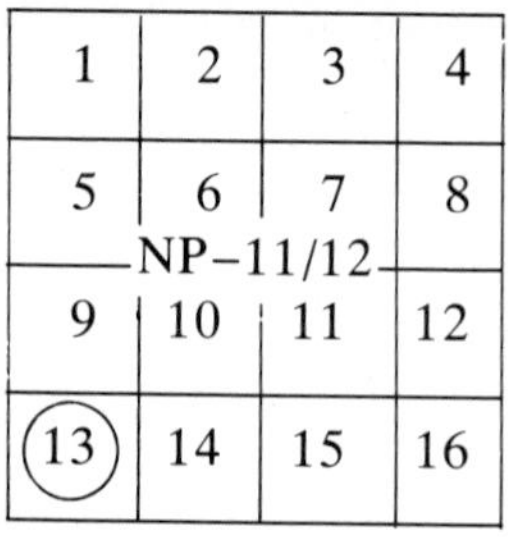

1	2	3	4
5	6	7	8
	NP–11/12		
9	10	11	12
(13)	14	15	16

North of 72° the convergence of the meridians necessitates a return to the 12-sheet division. Table 15 which lists the Canadian JOG series (in Chapter 8), illustrates the use of this numbering system.

The UTM Reference Grid

All Canadian topographic maps at scales of 1:250,000 and larger are drawn on the UTM projection and all except the 1:125,000 series carry UTM grid lines. It is not possible here to describe fully the UTM projection, but a brief outline is necessary to explain the grid lines and numbers that are found on the maps.

The UTM projection is conformal, which means that features drawn on the maps retain their true shape, and in a given area the scale of the map is constant in all directions. So that the distortion that is inevitable when a curved surface is projected onto a flat surface is kept within reasonable limits the projecting is done by zones which are 6° of longitude wide and which run from Pole to Pole as shown in Figure 39. Canada, as shown in Figure 40, is covered by Zones 7 to 22 inclusive. Once the area within the zone is flattened to the paper, a rectangular grid is superimposed, as shown in Figure 41. This is a metric grid with the point of origin being on the Equator 500,000 metres to the west of the central meridian of the zone. This is done so that all grid numbers will be positive. On Canadian maps the full measurement from the point of origin to the most western and southern grid lines on the sheet is given in the margin at the southwest corner of the map. All other grid-line numbers

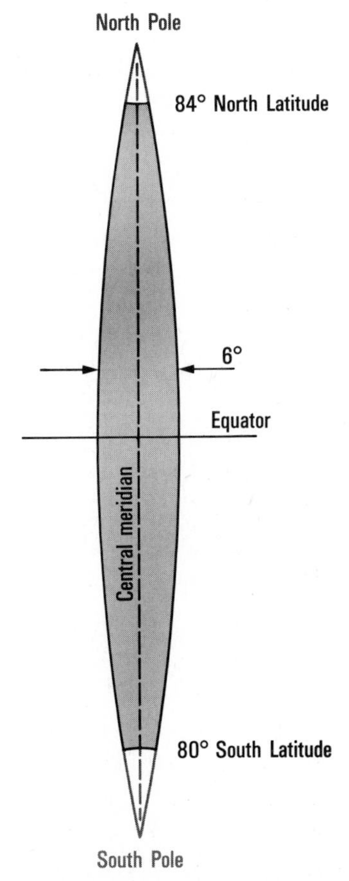

Figure 39 Shape of the UTM zone

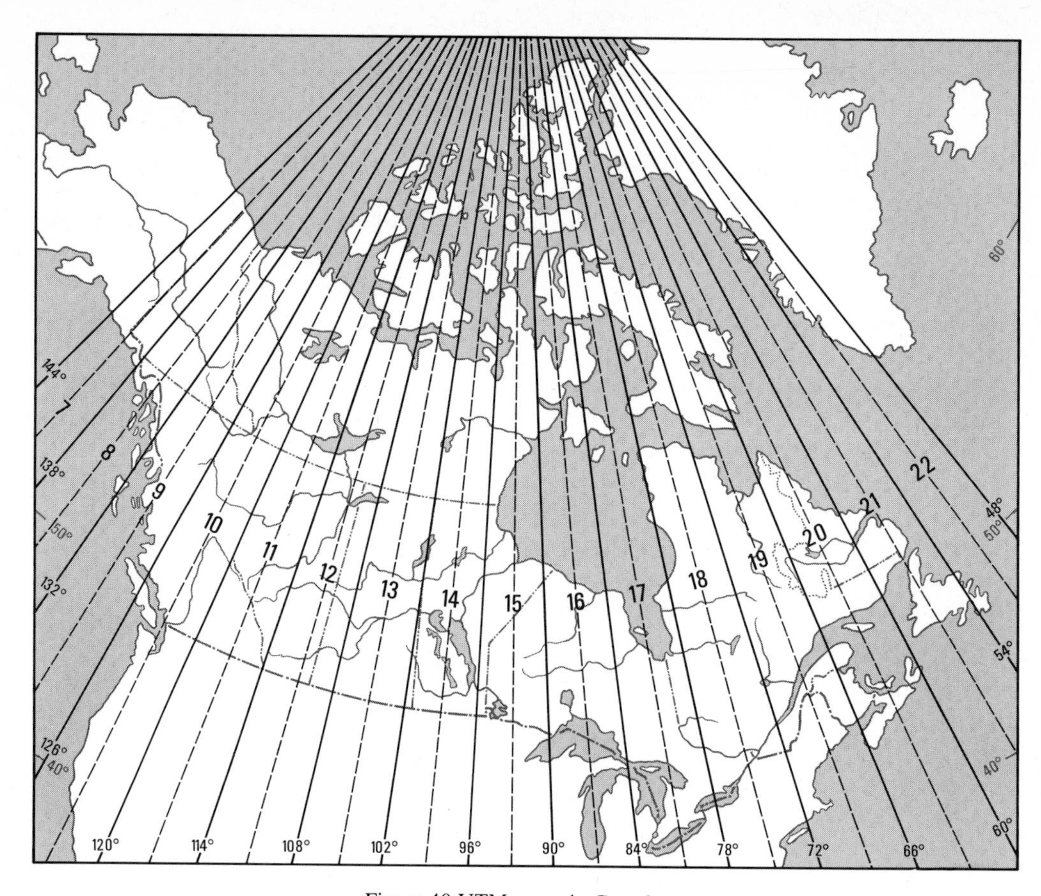

Figure 40 UTM zones in Canada

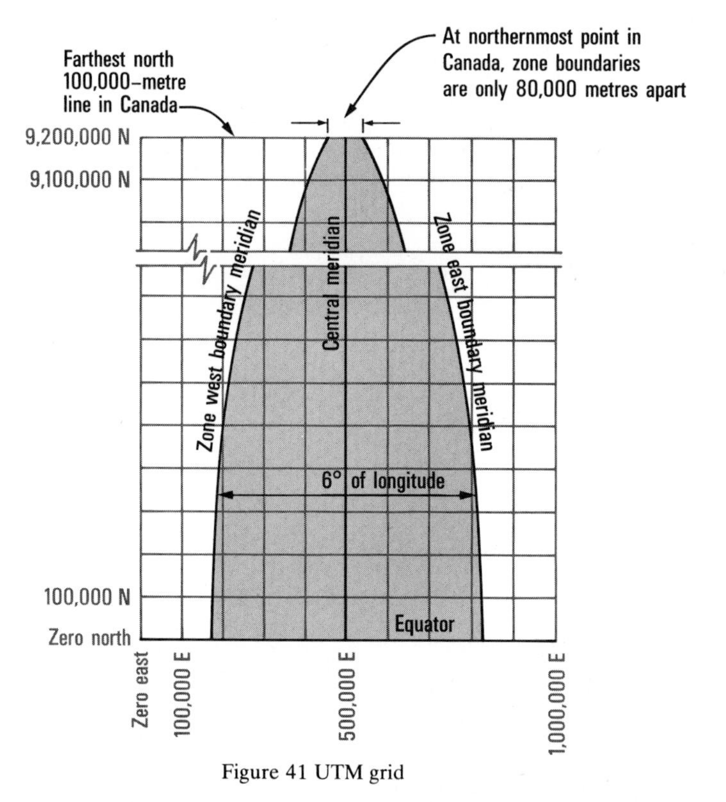

Figure 41 UTM grid

are limited to two digits for the 1:25,000 and 1:50,000 series and a single digit for the 1:250,000 series. As the grid system is the same for all series, the identical area of terrain is contained within the 1000-metre grid squares at the 1:25,000 and 1:50,000 scales that are identified by the same grid numbers. The 10,000-metre squares of the 1:250,000 series contain the terrain of 100 grid squares (10 by 10) of the 1:50,000 series, but of course the terrain depiction is much generalized due to the reduction in scale.

Military requirements demand that the grid of any zone be extended into both neighbouring zones to the extent of one 1:50,000 map sheet. This is done on coloured maps by

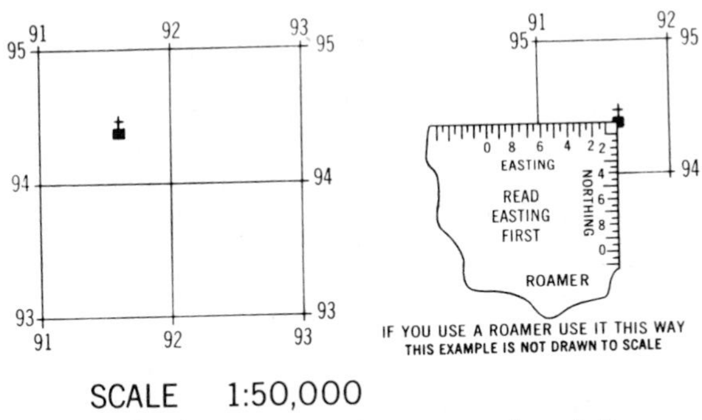

To find Map reference of ⛪ proceed as follows

1. Find Number of Grid Line West of ⛪ (91)
Ascertain number of tenths ⛪ is east of (91)
This is observed to be 6.
Set it down thus, 916. This is known as EASTING

2. Find Number of Grid Line South of ⛪ (94)
Ascertain number of tenths ⛪ is North of (94)
This is observed to be 4.
Set it down thus, 944. This is known as NORTHING

The Map reference of ⛪ is therefore 916944

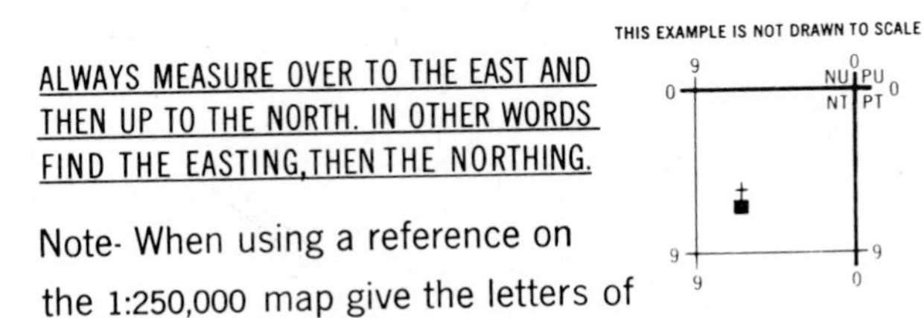

Map reference on 1:250,000 scale is NT9393

Figure 42 Method of finding a map reference

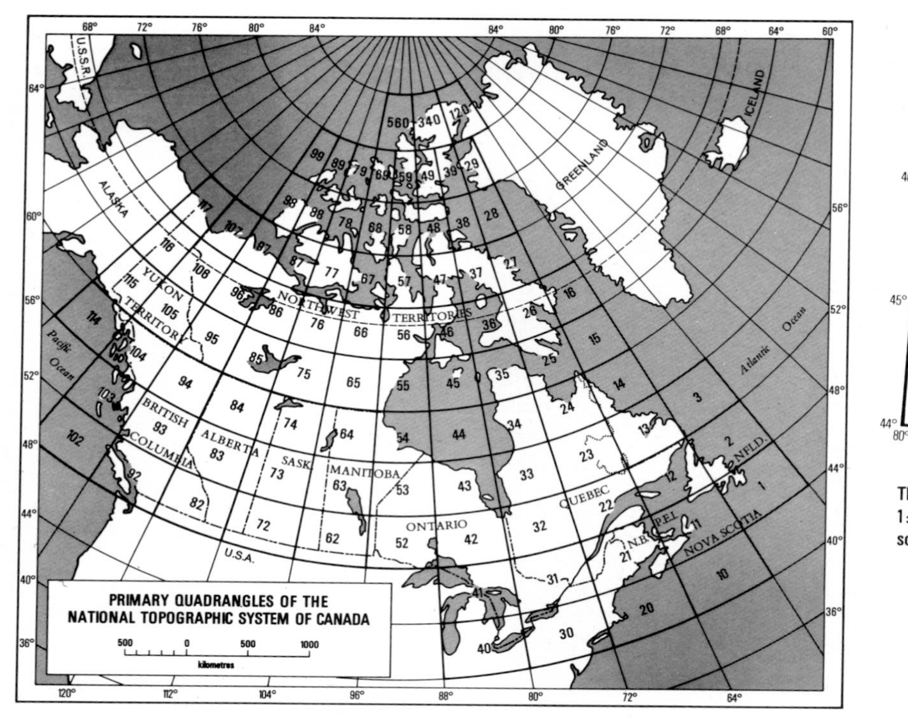

Figure 43 The primary quadrangles of the NTS system

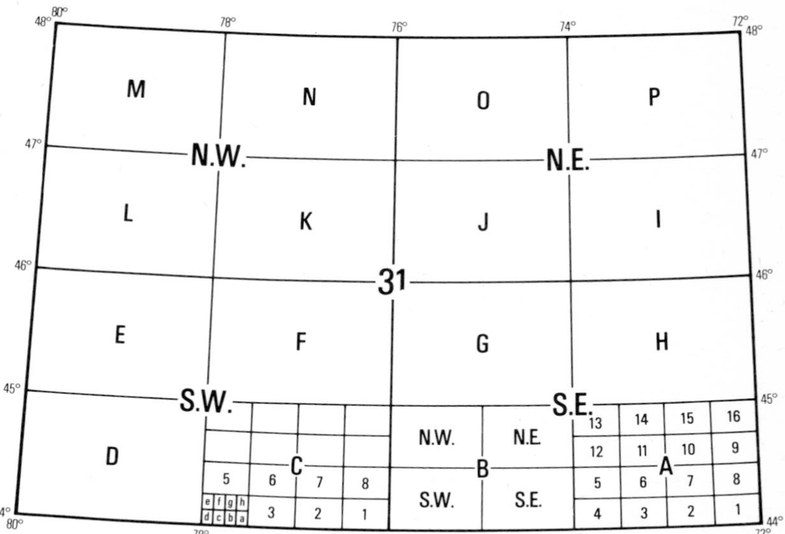

The national Topographic System of Canada includes the following scales: 1:1,000,000, 1:500,000, 1:250,000, 1:125,000, 1:50,000 and 1:25,000. Canada is covered by primary quadrangles 4° north to south and 8° east to west (16° east to west, north of 80° north latitude).

31 ______ A primary quadrangle of the 1:1,000,000 series.

31 S.W. ______ A quarter of the primary quadrangle is a sheet of the 1:500,000 series.

31 G ______ A quarter of a 1:500,000 sheet is a sheet of the 1:250,000 series. South of 68° these sheets are lettered A to P. North of 68° they are lettered A to H.

31 B / N.E. ______ A quarter of a 1:250,000 sheet is a sheet of the 1:125,000 series.

31 A / 8 ______ A sixteenth of a 1:250,000 sheet is a sheet of the 1:50,000 series. These sheets are sometimes published in east and west halves.

31 C / 4a ______ An eighth of a 1:50,000 sheet is a sheet of the 1:25,000 series.

Figure 44 The breakdown of a primary quadrangle

the use of brown grid ticks in the margin and on monochrome sheets by black ticks. Should trigonometric measurements be necessary over a zone boundary (as for example during artillery practice) the overlapping grid can easily be ruled on the adjoining sheet in pencil.

An explanation of how to make a grid reference of a point on a map is printed in the margin of every sheet. In brief, the grid line to the west of the point is read first, and the tenths of the distance eastward toward the next line is estimated. At the 1:50,000 scale this gives three digits which are termed the eastings of the point. The northings are read in the same manner starting with the grid line immediately to the south of the points. The grid reference consists of six digits. As such, a reference repeats every 100 kilometres, and if there is a risk of ambiguity the two blue 100,000-metre square identification letters, printed on the face of the map, should be included in the grid reference. Figure 42 illustrates the manner of making grid references.

16. Map Printing Methods and Map Accuracies

Map Printing Methods

The mainstay of Canadian map printing has always been lithography, a printing method discovered and developed by a Bavarian, Alois Senefelder, in 1798.[1] In essence it is based on the principle that if a mark is made on a smooth piece of limestone with a certain type of grease pencil, and if the surface of the limestone is kept wet, only the greasy line will pick up ink when the limestone is rolled with an ink covered roller. With the ink adhering only to the marked part of the stone, it is a simple matter to transfer the inked image to paper.

From this simple process grew the adaptable and versatile system of lithographic printing. Originally, maps that were to be lithographed were drawn directly on stones, but as this required the image to be drawn in reverse it was not popular. Transfer paper, a strong, grained paper coated with a thin layer of size which absorbs printing ink, was used at times so that the draftsman could draw his image right-reading and then transfer it to the stone in a special transfer press.

Copper plate engraving is an elegant method of printing the linework of maps, but it was considered too expensive for all but a few prestige products. A compromise between the elegance and expense of copper plate printing and the economy of lithography was achieved by engraving the map image, colour by colour, on copper plates and then transferring the image to stone for printing.

A photomechanical method of preparing printing plates called photozincography was developed at the Ordnance Survey in Southampton, England between 1855 and 1860. In this system the map image is photographed onto thin sheets of zinc that have been given a fine grain and coated with light-sensitive chemicals. The image is recorded on the zinc in the manner of the tin-type photography of the last century. The image is then 'worked up' by the pressman using special chemicals to strengthen the image and make it receptive to ink. The plate was then used directly on either a flat-bed or rotary press. The maps of the

Fortification Survey (Chapter 2) were printed by photozincography, and were almost certainly the first Canadian maps to be printed by this method.

A further development in photozincography was the making of a photographic negative, on glass, of the map manuscript. The negative was then held against the treated zinc plate and the image transferred to the zinc by shining arc lights through the negative. In essence this system is in use today though stable base photographic film has replaced the glass negatives.

It is difficult to be certain which system of printing was used on official Canadian maps in the nineteenth and early twentieth centuries as various methods were available in the print shop of the Department of the Interior. Often the method used depended on which press was available rather than which system was most appropriate for the map at hand. Maps of a series that were likely to be reprinted were usually engraved on copper, but transferred to stone for printing. Originally all of the map work was done in the Department of the Interior, but early in the twentieth century much of the printing was done by the Government Printing Bureau or put out to contract by them. In 1912 the Department of the Militia and Defence obtained its own printing equipment.

The annual reports of the Lithographic Office of the Department of the Interior record the advance of printing techniques by the Canadian government. For example in 1904 we find the following comment:

> During the last ten or twelve years, photozincography has been much improved, and is now used almost exclusively for work like ours in the Ordnance Survey office at Southampton, in the Survey of India Department and in all the great survey offices everywhere. The plans are photographed on thin zinc sheets and printed from the zinc instead of being transferred to and printed from stones. The process is far more rapid and the work much better than with other methods depending upon photography and lithography. Preparations are being made for the introduction of the process here. The necessary apparatus and materials have been ordered and everything will be ready in time to handle the plans of the present season's surveys when they are complete for printing.

The report of the following year states that the photozincography equipment was in operation. In the report of 1911 the following advantages of photozincography are mentioned:

> A new frame for hanging the copying camera and copying board has been installed; it is perfectly rigid and provided with means of adjustment both for the copying board and for the camera. A new graduation has been made for setting the focus to enlarge or reduce to any scale: it is very accurate.
>
> Photozincographs in colours are now made from a single negative from which as many zinc transfers are made as there are colours to be printed. For each transfer, all the lines which must not show are painted out on the negative and after the transfer is made, the paint is washed off. The process is repeated for each colour. All the transfers being made from one negative, perfect registration is obtained.
>
> There were 196 wet plate negatives and 118 photozinc transfers more than last year.

The fact that some drawing on stone was till done in 1913 is brought out in the report of that year:

> A power paper-cutter was added to the equipment which now consists of a flat-bed power-press, a rotary offset power-press, three hand-presses for transferring, a zinc-plate graining machine and a power paper-cutter.

> The work is for the greater part photozincographic printing. A few maps are printed on stone; others are engraved and transferred.
>
> The staff of ten employees is the same as last year.

The report of 1914 gives a very good summary of the state of map printing at that time, and is reprinted here:

> The work of the lithographic office continues to increase steadily . . . Two power presses are used, one a flat-bed machine capable of printing either from stone or from zinc plates, the other a rotary offset press printing from zinc plates only. One essential difference between the presses is that for the flat-bed press the work on the plate or stone is reversed, so that when the sheet of paper to be printed comes in contact with the work and receives a print, the print reads correctly. With the offset press the work on the plate reads correctly; a reverse print is made to a rubber blanket which in turn prints a correct copy on the paper. It is therefore necessary in preparing work for these presses to keep in mind this difference, as any plate prepared for one press must be reversed before the other press can print it. This reversing may be done by transferring but this usually thickens up the work and causes a loss of sharpness. A preferable way is to reverse by photography, and as most of the work is photographed, it is merely necessary to determine upon which press a job is to be printed, and the photographer arranges the matter by copying either direct or through a mirror as desired. The offset press is a later development in lithography and is capable of a higher rate of speed than the flat-bed. It is therefore used in long runs. The printing of annual report maps and of township plans forms the bulk of the work turned out. Of the latter, 203 copies only are printed; 3 on linen, 170 on thin paper for mailing purposes, and 30 on thick paper for ordinary office use.
>
> More time is required for preparation and adjustment of press to meet the conditions relative to the printing on each kind of paper than would be required for a straight run on only one kind of paper. The same would apply regarding the necessity for frequent changes when more than one colour is used in printing a plan or map.
>
> The flat-bed press is easier to change from one colour to another, and is consequently used much for colour work or for short runs. The printing of the 3-mile sectional maps in three colours, black, blue, and brown, has been undertaken and provides considerable additional work for the flat-bed press. Reprints of township plans originally issued in colours have also given much colour work.
>
> The largest size of paper used is 24 inches by 34 inches, so that the maximum size of map which can be printed is about 22 inches by 32 inches, varying a little with the allowance for margin.

Gradually the flat-bed presses were taken out of service and by World War II only rotary offset presses were in use. The great advance in post-war printing techniques was the development of negative scribing in which the map manuscript was printed photographically on plastic sheets that had been coated with an opaque paint. Using a scribing tool the draftsman removes the surface coating of all lines of a given map colour, thus making an artificial negative. Aluminium has replaced zinc as the press-plate material, but in other respects the lithographic printing system is much the same as in pre-war years.

Provincial government mapping agencies have always had their map printing done by contract. The printing methods of the private sector have always been much the same as those of the federal government.

The Accuracy of Canadian Topographic Maps

There are three distinct phases in producing a topographic map, and it is inevitable that certain errors may be introduced into the map at each one of them. The first phase is the surveying of the framework of the map. This is usually referred to as the control survey. The second phase is the drawing of the map manuscript based on the control points. The final phase is the fair drawing of the manuscript and its printing. In Canadian mapping practice all three phases have gone through a series of changes which have improved the accuracy and volume of map production. In examining these changes it is convenient to consider three distinct periods, namely the pre-World War II period, the post-war period up to about 1960 and the modern period.

Before World War II the accuracy of the maps varied widely, and generally speaking the only maps that had any pretense to accuracy were those of the urban and farming areas of eastern Canada. Even in settled areas the accuracy was low when judged by European standards. In smaller, well-developed countries the topographic surveys were always tied to fundamental surveys of high precision. When topographic surveys commenced in Canada in the early days of the present century, no geodetic triangulation stations existed. As late as 1927 the Department of Interior's *Manual for Topographers*[2] included the following statement:

> The ideal method of providing main control is by triangulation because no other method has the same facility for spreading the control consistently to cover area. Unfortunately, however, conditions in Canada generally do not lend themselves to the extension of small triangles. For facility in this work one requires local eminences with clear vision between them. In the great western areas of smooth prairie the local eminences are lacking and in eastern Canada they are generally covered with a tree growth, through or above which vision cannot be obtained within allowable limits of cost.

It follows that in Canada traversing, i.e. the measuring of distances and angles along a given route, was the normal method of establishing the framework of maps. Traverses can be done accurately or they can be done hurriedly, and unfortunately the latter was often the Canadian practice. In the original survey of the Prairies, which later provided the framework for the maps of the three-mile series (Chapter 3), the work was carried forward at top speed. It had to be, because the settlers were pouring onto the Prairies and it was government policy to provide each family with a surveyed farm lot. Later the Canadian Pacific and other railway companies demanded property surveys immediately after they had selected the railway right-of-way. In some cases they threatened to carry out their own surveys if the government did not provide them. The latter option was of course not acceptable, but the pressure on the government survey establishment was severe. In 1921, when the conditions on the Prairies had stabilized, the Department of the Interior had one of their senior engineers, J. B. Milliken, investigate the errors in the basic surveys that formed the control for the three-mile series. The errors were so great that it was felt that the Prairie landowners would be seriously disturbed by the information in the report, and although it was used within government offices, the Milliken Report was not made public for 50 years.

But before one is critical of the early Prairie surveyors, one must remember the urgency of the time and the fact that most of the work was done by contract. In his report Milliken gives this view on the efficacy of contract surveying:[3]

> This system seemed at the outset the best one to adopt. It was one of the features borrowed from the United States. The scale of prices per mile was at that time fixed in accordance with the nature of the survey and in anticipation of the character of the difficulties expected to be met with thereon. During the interval 1869 to 1872 surveyors of block outlines found that the prices per mile previously fixed on were in some cases not sufficient. At the time when the prices for contracts

> of various kinds were fixed, the nature of the country through which these lines were to be run was not sufficiently well known. Surveyors afterward making these surveys at their own personal expense found that the country was not nearly so uniform in character as had been anticipated when the prices were fixed, and also that a much smaller proportion than had previously been expected was open prairie. Other difficulties previously unknown impeded their progress. Extensive marshes, thick woods, windfalls, and unfavourable weather all contributed to delays. Where the conditions were favourable the surveyor was impelled to hurry the work as he was continually impressed with the possibility that what followed might absorb more than a fair proportion of time, and thus cause him loss.

Although after 1872 the survey of block outlines (the perimeter of 16 townships, four in latitude and four in longitude) was done by salaried surveyors, the interior lines of the block continued to be done by contract until 1917.[4]

Many serious errors escaped the investigations of Mr Milliken and were not discovered until the 1:50,000 mapping of the Prairies was undertaken after World War II. Some use was made of the original traverses in this new mapping, but to bring the old surveys to a common datum and to correct discrepancies it was found necessary in many cases to resurvey the area using electronic distance-measuring techniques. The original three-mile maps were not recompiled even though the Milliken Report revealed that appreciable errors existed in the mapping.

In the settled parts of eastern Canada, before World War II, the methodical traverses along country roads formed the normal method of framework survey. The more leisurely pace, and more frequent inspection of work, provided the conditions for greater accuracy. But these were still secondary surveys that were not tied into first-order control. M. F. Phelan, for many years in charge of control surveys for the Geographical Section of the General Staff, makes the following statement about pre-World War II map accuracies:

> The roads, with features beside them, were quite accurately placed. Any considerable distance from the roads, especially in wooded, rough country, the contouring, etc. was frequently faulty. After the fixing of geodetic points by the Geodetic Survey at a later date, it was sometimes slightly embarrassing to find that the plotted point was on the side of the hill instead of on the top. The vagaries of water courses beyond open country could not be relied on, but on the whole there was little adverse criticism by those who had occasion to use these sheets.[5]

Phelan of course describes maps compiled by the plane-table method based on the control traverses that he and his soldier surveyors ran through the area to be mapped. At the same time as these surveys were being conducted in eastern Canada the phototopographic surveys, described in Chapter 2, were going ahead in the mountains of British Columbia. These were even less accurate than the plane-table methods, but the area being mapped was mostly uninhabited, and accurate mapping was not needed.

In the 1930s, selected areas of northern Canada were being mapped by the air-oblique method based either on astro-fixes or, at best, range-finder traverses down the larger river courses (Chapter 8). By European standards this procedure would be classed as sketch mapping.

The establishment of a reliable vertical datum has always been a problem in Canadian mapping. On the earliest of topographic surveys, such as the Fortification Surveys, the elevations and contour values were based on the water-level of the St Lawrence River in the vicinity of the area being surveyed. These local datums were sufficient for the time but were not suited for a large map series.

When the military topographic surveys were commenced in 1904, the vertical datum was mean sea level at New York City.[6] This datum had been extended by US engineers to their cities along the St Lawrence River and the Great Lakes. The Canadian military surveyors evidently saw nothing anomalous in obtaining survey data from the potential enemy!

On the Prairies and in the western mountains the elevations above sea level were originally obtained from the railway surveys. When the three-mile maps were converted from the Old to the New Style in the 1920s most of the contouring was done simply by running barometer lines between points where the railways crossed the township roads. In the 1950s the first editions of the 1:50,000 maps of the same area were surveyed in the vertical sense by running level lines of very modest accuracy (errors were kept within 12 inches per mile),[7] again between points on railway lines.

In due course the Geodetic Survey ran precise level lines along a dense network of roads and railways. Tide gauges have been established at numerous points on the three oceans and in the Arctic islands. Today all topographic maps are tied to these precise surveys, and a comparison of elevations on early and later editions of many topographic maps reveal serious errors in the earlier work.

World War II brought changes to the whole concept of mapping Canada. A number of Canadians were trained in photogrammetric mapping methods, and many of them wanted to stay in the mapping business after the war. As has been mentioned, pilots and photographic aircraft were available, and by 1947 the Canadian air survey industry was established and thriving. Government mapping offices were also being modernized by the purchase of new equipment. Despite this modernization the accuracy of Canadian topographic mapping did not greatly improve in the immediate post-war years. Photogrammetric methods were introduced, but the control surveyors were still faced with pre-war survey techniques which left much to be desired when the work progressed into difficult country. The Geodetic Survey could not work quickly enough to stay ahead of the topographical surveyors.

In the horizontal sense, the breakthrough which vastly improved the accuracy of the positions of features on Canadian maps came in the early 1960s with the adoption of electronic distance measuring devices (EDM's) and the utilization of large electronic computers. The EDM's were used in the basic triangulation where all sides and all angles of each triangle were measured to give a very accurate value for the positions of the triangulation stations. EDM's were also used in secondary traverses which were run through the area being mapped. Because long lines could be measured with great accuracy, these secondary surveys were infinitely stronger than the traditional road traverses of the previous four decades.

The electronic computers were used in the photogrammetric process called a block adjustment. In such work a large area embracing 20 or so maps is surveyed as a block. An EDM traverse is run around the perimeter, and the position of each station of this traverse is carefully identified on the aerial photography covering the area. This photography is flown in lines with each photograph in a line overlapping its neighbour by about 60% and each line overlapping the next by about 30% (see Figure 43). A number of points of ground detail (a distinctive rock, a stream junction, etc.) are chosen on the photographs in the parts of the photos that overlap the adjacent lines of photography. These 'tie points' are measured very precisely on a grid coordinate system provided for each photograph, but as each point is chosen so that it falls on three or more photographs, the individual photo coordinates can be combined into a grid system covering the whole block. The electronic computer is used to effect this joining of photo coordinates into a common grid, and it is

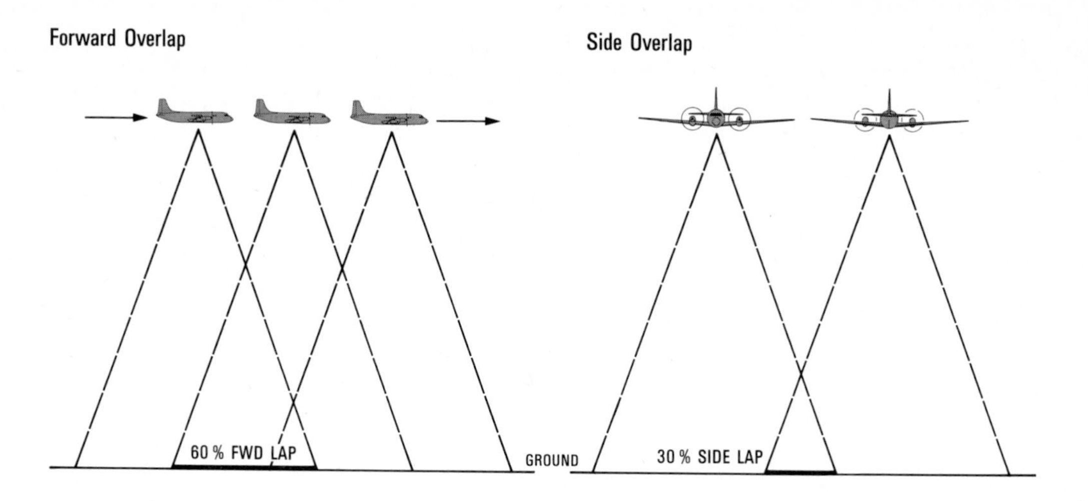

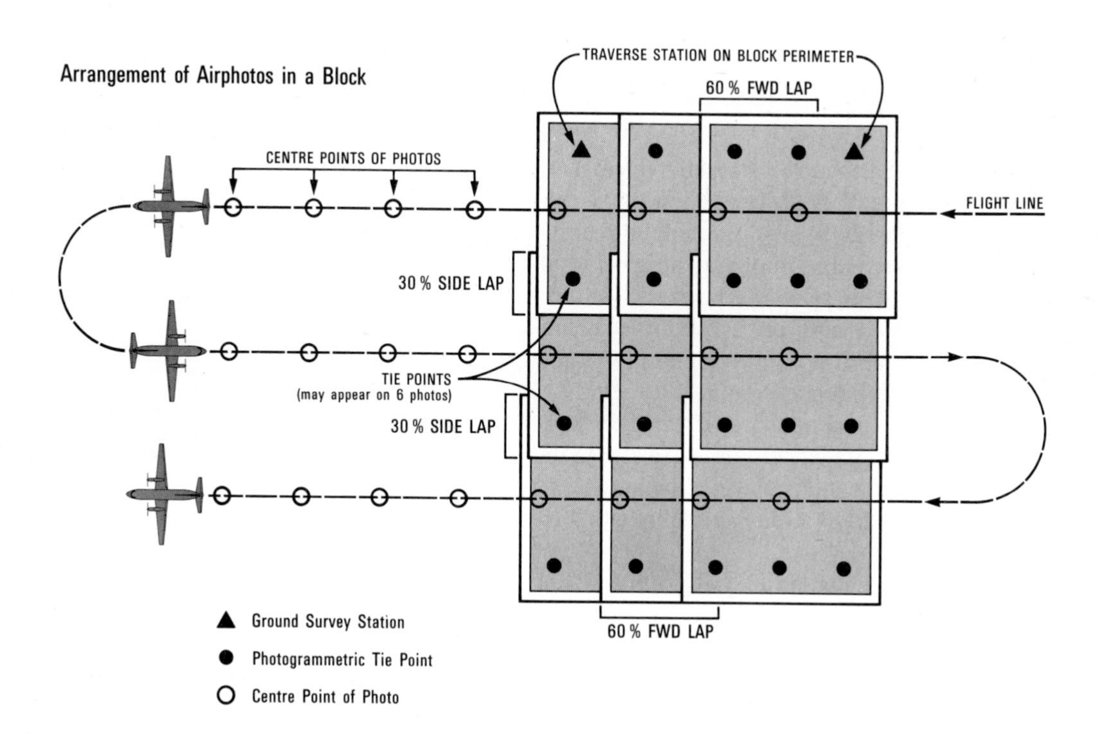

Figure 45 Aerial photography in a block adjustment

also used to adjust the common grid to the basic UTM grid of the ground survey.[8] It was only with the adoption of the block adjustment system that a true statistical assessment of the accuracy of Canadian topographic maps could be made.

In countries with large survey establishments, the usual practice is to test by field surveys the accuracy of a substantial number of maps. In Canada relatively few maps are checked in this way, but on the basis of the results of these few tests, an estimation of the accuracy of all other sheets is made by examining the layout and accuracy of the control surveys, noting the instruments used in the compilation of the maps and the scale and quality of the aerial photography. It is of course not as satisfactory as a more rigorous method based on a large sample, but as the resources spent in testing cut into the resources available for mapping, this extrapolation from a small sample must suffice.

In Canada, as in most other countries, the accuracy of the topographic mapping is assessed for two aspects of the map, namely the accuracy of the positions of features and the accuracy of the elevations shown. In making both assessments the accuracy is judged with relation to the most precise horizontal and vertical geodetic surveys in the region. It is the practice to express the horizontal accuracy of a map in terms of the number of metres that well-defined points might be found displaced from their true position on the map if a test survey were conducted in the area. Indefinite points, such as the edge of a forest or a bend in a river, would not be chosen for the test, and neither would other features that have been intentionally displaced to improve the clarity of the map. For example, on most topographic maps the roads are shown much wider than they actually are, and this in turn means that roadside buildings are displaced away fron the centreline of the road. A corner of such a building would therefore not be suitable for use in testing the map.

In testing the vertical accuracy it is the usual practice to select a number of well-defined points and read off their elevations by interpolating between contours. These points are then surveyed by field methods from bench marks established by the Geodetic Survey.

In assessing the results of these tests it is realized that variations will exist in the size of the errors and unless something is seriously wrong with the mapping small errors will occur more frequently than large errors. With these facts in mind it is usual to state the accuracy of the map in terms of a limit of displacement that is not exceeded by 90% of the test points. The following are the horizontal and vertical standards for Canadian maps.[9]

Horizontal Standards

Scale of map	Rating			
	A	B	C	D
1:25,000	12.5 metres	25 metres	See Note 1	See Note 2
1:50,000	25 metres	50 metres		
1:100,000	50 metres	100 metres		
1:200,000	100 metres	200 metres		
1:250,000	125 metres	250 metres		
1:500,000	250 metres	500 metres		

Notes:
1. Will not meet serials A and B.
2. Sketch map or uncontrolled mosaic.

Vertical Standards

Rating	Description
1	Half the contour interval used on the map.
2	One contour interval used on the map.
3	Will not meet the standard of rating 1 or 2.
4	An uncontoured map.

Note: Map ratings are given with the horizontal rating preceding the vertical. For example, a map with a horizontal rating of B and a vertical rating of 2 would be listed with accuracy B2.

Although it is impossible to give a complete listing of the accuracies of all published topographic maps, some general statements can be made on the subject.[10] The 1:25,000 series covers areas that are easily accessible, and the field surveys for this series were carefully done. Most maps of this series have a rating of A1 or A2.

The 1:50,000 series includes maps with ratings from A1 to C3. In areas covered by the 1:25,000 series the maps are generally of A1 category by virtue of being derived from the larger scale. The older maps of the series, i.e. those surveyed between the end of World War II and the introduction of mathematical block adjustments (circa 1962) are generally of low category. B2 and B3 ratings predominate, but some C3 sheets exist. The newer sheets south of the Wilderness Line are almost all A1 or A2.

As the accuracy of topographic mapping in the north is in direct proportion to the funds spent on field survey, mapping of lower accuracy was planned for sheets north of the Wilderness Line where great precision is not a requirement. As a general rule, sheets falling on the mainland but north of the Wilderness Line are rated A2, while sheets of the Arctic islands are generally B2.

The 1:125,000 series contains a 'mixed bag' of high and low ratings with the latter predominating. This is the result of using map bases that were surveyed many years ago. No concerted effort has been made to improve the topographic bases in this series as precise measurements are rarely taken from 1:125,000 maps.

The 1:250,000 series is another series with a wide range of accuracy ratings. Those sheets that have derived from 1:50,000 coverage are almost all of A1 category. The remaining, all in the north, are in general rated B2 and B3.

The Surveys and Mapping Branch has initiated a long-term programme to resurvey and recompile all substandard maps of the 1:50,000 series, and to derive accurate maps at the smaller scales from this new production. Some 15 to 20 years will elapse before this work is completed.

The progress in improving the accuracy of Canadian maps continues. Spectacular advances have been made in the development of equipment for determining the position of points on the surface of the earth. Use is made of special survey satellites which rotate around the earth on carefully measured orbits. Signals are sent out from the satellites and by measuring the Doppler Shift[11] of these signals as the satellites pass overhead, the position of the station on the ground can be computed. Between these Doppler stations, intermediate stations are surveyed by inertial survey equipment which measure with incredible precision the movement of the vehicle, helicopter or truck, carrying the inertial equipment. Both the Doppler and inertial survey equipment give results to a fraction of a metre in position and elevation. The electronic distance measuring surveys, which 20 years ago provided the breakthrough into the era of accurate mapping, are now referred to (with some scorn) as 'traditional' surveys by young surveyors who have quickly adapted to space-age technology.

17. Geographical Names on Canadian Maps

Geographical names are as fundamental and as necessary to a country as family names are to the people that inhabit it. All nations have named the physical features of their land (but not necessarily all of them), and almost all countries consider this naming a serious matter worthy of government control. Geographical names are very much a part of a country's history; they perpetuate the identity and character of the earliest inhabitants, and in the newer countries they express the hopes and expectations of the explorers and settlers who rediscovered the land. This is particularly important in Canada which is a mosaic of many ethnic groups, and although the sharp edges of the pieces of the 'mosaic' are today less distinct than 100 years ago, the pattern of early settlement can still be seen. The geographical names are most important in preserving this regional heritage.

A brief study of the names on a map of Canada reveals this close connection between place names and early settlement. Consider the jaunty sea-going names along the Newfoundland coast and the Irish family names that predominate inland. Consider, too, the stern Scottish names of Nova Scotia. The place names of Quebec are of course mostly French, but are also deeply religious. In Ontario there is a mixture of Anglo-Saxon and German names dating from the days following the American Revolution when both British Loyalists and Hessian soldiers settled in the wilderness. The Prairie provinces were explored by both French and English to extend the fur trade, but, except for the establishment of a few settlements in the vicinity of the trading posts, the fur traders did nothing to encourage the occupation of the land. The great influx onto the Prairies came with the railways; and it came with a rush. Villages were established so rapidly along the railway lines that we find the settlements often named for convenience in alphabetical order. Thus, going west from Portage la Prairie we find Arona, Barr, Caye, Deer, Exira, Ferdale, and so on toward Calgary. British Columbia was explored from the Pacific Ocean by British, Spanish and Russian navigators, with each nationality leaving evidence of its passage in the names along the coast. Explorers such as Alexander Mackenzie, John Stewart, Simon

Fraser and David Thompson came over the mountains from the Prairies to the coast. En route they named certain prominent features, and in turn at a later date had features named in their honour. But the extensive naming of the province had to wait for the arrival of British administrators who came during Queen Victoria's reign. The Royal Family was honoured by names such as Prince Rupert, Queen Charlotte, Princess Royal and Victoria. Heads of departments in the British government, naval and military officers and other dignitaries were also 'immortalized' in this way.

Even though Canada is a mosaic of many cultures there is, in Canadian place names, a unifying thread, and this is the interesting and often beautiful family of native names. Many of these names now describe places of primary importance such as Toronto, Winnipeg, Saskatoon and Kamloops. They also have been given to extensive regions such as the Muskoka Lakes, the Torngat Mountains and the District of Keewatin.

The primary use of place names is, of course, to provide a ready means of reference to the geographical features of the country. The map-user may not be too concerned with the historical and ethnic aspects of place names, but he has certain definite requirements. He wants names he can pronounce without too much difficulty, he wants a sufficient number of features to be named so that he can talk about, or write about, the land without ambiguity, yet he does not want his map to be so cluttered with names that he has difficulty determining which name applies to which feature. He does not want two or more features on his map to have the same name. Above all, he wants the names on his map to be the same as those used by the inhabitants of the area shown on the map.

The Canadian Permanent Committee on Geographical Names

With so much history embodied in place names, and yet with a set of very definite specifications regarding names presented by the map-using public, the regulation of a country's geographic names must be placed in the hands of specialists. In Canada these specialists are embodied in the Canadian Permanent Committee on Geographical Names (CPCGN). The CPCGN started life in 1897 as the Geographic Board of Canada. From the first it was charged with the task of recording and standardizing the place names of Canada. In 1948 the Board was renamed the Canadian Board on Geographical Names and in 1961 it was reorganized as the Canadian Permanent Committee on Geographical Names. The Committee is composed of a representative from each of the provinces and from each of the various federal government departments that have an interest in mapping. The Department of Indian and Northern Affairs represents the Yukon and the Northwest Territories. A secretariat is provided by the Department of Energy, Mines and Resources.

To ensure that uniform standards and principles of naming are followed in all parts of Canada, a set of 'principles of nomenclature' have been evolved. These are as follows[1]:

1 Names created by legislation are accepted by the Committee.
2 Postal authorities, railway companies and resource development companies are encouraged to seek the advice of the Committee concerning names connected with their operations.
3 First consideration is given to names with well-established public usage. Unless there are good reasons to the contrary, this principle should prevail.
4 Established names that have proved acceptable and satisfactory should not be changed or altered.

5 Names applying to various service facilities (e.g. post offices, stations) should conform with the names of the communities which they serve. Only a single name should be given for each feature.
6 Duplication of names to the extent that it may cause confusion should be avoided.
7 Approval of both an English and a French form of a name for the same feature may be sanctioned only where both have well-established usage.
8 The application of a personal name is not made unless it can be demonstrated that exceptional circumstances exist and it is in the public interest to do so.
9 Preferred sources for names are: descriptive names appropriate to the features; pioneers, war casualties and historical events connected with the area; names from native languages currently or formerly identified with the general area.
10 Indian and Eskimo names shall be recorded according to the best recognized orthography.
11 Names should be euphonious and in good taste.
12 Qualifying words such as upper, new, west branch, and big may be used in an area to distinguish between two or more features with identical names.
13 A generic should indicate the nature of the feature.
14 Foreign names shall conform to international regulations.

Some of the above principles have been broken in Canadian practice, but only at the expense of a certain amount of confusion. The whole basis of a workable place-name system is the permanency of names, yet Berlin, Ontario, was changed to Kitchener during the First World War for patriotic reasons. Castle Mountain was changed to Mount Eisenhower to honour a war hero. Fortunately such occurrences are rare. The rule discouraging personal names is sound, but it was not always followed in Canada. Many quite minor government employees have been honoured in this way. For example, the gold mining town of Kirkland Lake, Ontario, was named after Winefred Kirkland who was for many years secretary to the Surveyor-General of Ontario. Patrons of exploratory expeditions into the Canadian Arctic were honoured to compensate for the funds they donated; thus we find on the Canadian maps the names of Felix Booth, the London distiller, and the Ringnes brothers (Amund and Ellef) both Norwegian brewers.

Although not a part of the official principles of nomenclature there are two unofficial rules that, because they are eminently sensible, are usually followed. The Committee insists that a uniform spelling of names be maintained in a given area. A few mistakes have slipped through in the past to cause uncertainty. Chilliwack City and Chilliwhack District Municipality, both in the same area of British Columbia, are cases in point. The second unofficial rule is that descriptive names, such as Lion Head Point or Granite Mountain, must accurately describe the feature. A developer of cottage properties may want to call a weed-filled slough 'Clearwater Lake', but this is not condoned.

The Gazetteer of Canada

There are two types of documents that provide a permanent record of a geographic name: the topographic map of the area on which the name appears, and the official gazetteer in which the name is listed and described. Topographic maps need no further discussion here, but the Canadian set of gazetteers is worthy of a short description and review.

The forerunners of the official Canadian gazetteers were lists of names approved by the Geographic Board of Canada. But from time to time their publications included alphabetical lists of names, usually by political units, with a short description of the feature named and the origin of the name. Thus the report of the Board for 1910 included 'Place Names in Quebec'; 'Place Names—Thousand Islands St Lawrence River' and 'Place Names in Northern Canada' (the then Northwest Territories and Yukon). Later the Board published separate volumes of such information with the addition of the geographical position of the features named. *Place Names of Prince Edward Island* appeared in 1925; *Place Names of Alberta* in 1928 and *Place Names of Manitoba* in 1933. But these appeared in an *ad hoc* fashion and in 1949 the Board drew up a plan for the publication of a national gazetteer in cooperation with the provincial authorities. The first volume *Gazetteer of Canada Southwestern Ontario* appeared in 1952. A separate volume has since been published for each province or territory.

In each volume the geographical names are listed alphabetically. Opposite each name there is an indication of the general location of the feature either by stating the number of the map on which it falls, its DLS township code, or its position in relation to well-known landmarks. Finally, and most important, the geographical position of the feature in latitude and longitude is given. Before 1967 the volumes were published in English only. Between 1967 and 1973 all the introductory material, such as notes explaining the use of the gazetteer, a list of abbreviations, principles of nomenclature, etc., was in both French and English, but the listings were in English only.

Volumes published since 1973, except the one for Ontario, have been completely bilingual, but at first glance the listings may not appear to be so, as the generic terms (lake, river, etc.) are printed only as they appear on the map. Thus the 'lake' in Horseshoe Lake would not be translated into French and Fer à Cheval, Lac would not appear in English. A glossary at the front of the text has been included in each volume since 1973. It gives a translation of all generic terms used.

Each of the twelve volumes making up the *Gazetteer of Canada* has its own peculiarities. The following is a short review of each.

Alberta

The current edition was published in 1974 and hence is completely bilingual. As mentioned before, the bilingual nature of the listing of names is not instantly apparent because all names including their generic terms are listed exactly as they appear on the topographic map. For example, Newton Creek is not translated, but Ninton is followed by the explanation (Hamlet-Hameau). The 1:250,000 map on which each place name falls is indicated by the sheet number. Following this the DLS township is indicated in the normal code; for example, Nipisi River is shown as being in 82-5-W5 which translates as Township 82, Range 5, west of the 5th meridian. The final column of the listing gives the geographic position of the name to the nearest minute of latitude and longitude. The glossary is somewhat disappointing as it does not appear to reflect the peculiarities of Prairie usage. For example, the eastern Canadian definition of a bluff is 'a certain elevation with one or more precipitous sides'. On the Prairies a bluff is a row of trees planted as a windbreak.

This volume has 153 pages of names listings.

British Columbia

The current edition, published in 1966, is the most unsatisfactory of all the volumes. The deficiency is in the geographical location of each feature which is given only to the nearest degree of latitude and longitude, followed by the quadrant designation NW, NE, SW or SE. This means, in terms of 1:50,000 sheet lines, that the feature being searched for may lie anywhere on the two maps covering the quarter of the degree quadrangle. The other deficiencies are minor by comparison. It is published in English only and has no glossary.

On the credit side, it does include a useful list of major rivers, lakes, islands and incorporated settlements in addition to the usual introductory material. Also included is a short history of the province.

The listing of names is contained on 739 pages. A new edition correcting the faults mentioned above is in preparation.

Manitoba

The current edition is dated 1968. The introductory material is in both French and English, but the listing of names is in English only. No glossary is included. The name listing includes a separate column defining the feature which in nine times out of ten is a repetition of the generic part of the name (as in Nelson Lake–Lake). This is followed by the location of the place in relation to a well-known settlement or topographic feature and the DLS township in which the name is found. The final column is the latitude and longitude of the named feature to the nearest minute. The volume has 93 pages of names listings.

New Brunswick

Published in 1972 this gazetteer is bilingual only in so far as the introductory material is concerned. It has an extensive and interesting glossary in which can be seen the influence of the Acadian French in some of the geographical terms such as: *barachois* (a body of water enclosed by a sandbar with a narrow entrance), and *batture* (a raised subsurface area of rock that causes continual breaking of waves). The listing consists of 213 pages.

Newfoundland and Labrador

Published in 1968 this is bilingual in so far as the preparatory material is concerned, but is English only in the listing of place names. No glossary is included which is unfortunate as the Newfoundland vocabulary is rich in unusual geographic terms. An unusual feature of this volume is the inclusion, on yellow paper, of a gazetteer of the French islands of St Pierre and Miquelon which lie just off the south shore of Newfoundland.

The Newfoundland and Labrador section is 252 pages long while that of St Pierre and Miquelon consists of 8 pages.

Northwest Territories

This volume, published in 1971, is still in its provisional edition. There is no prefatory material, and the listing consists of the place name, the type of feature (if this is not embodied in the name) and the geographical location to the nearest minute of latitude and longitude. It is in English only and consists of 73 pages each $8\frac{1}{2}$ by 11 inches (thus being in larger format than all other gazetteers which are uniformly $6\frac{1}{2}$ inches by $9\frac{3}{4}$ inches).

Ontario

The current gazetteer was published in 1974 but because of its great length (822 pages) it had been in work long before the decision was taken to make all gazetteers produced by the federal government bilingual. Consequently only the introductory material is in both French and English. There is no glossary, though one is certainly needed to explain those terms that have a special Ontario meaning, such as geographic township, municipal township, improvement district, Patricia Portion, etc. The text location of each feature is restricted to its township, county or district, whichever terms apply. The geographical reference is given to the nearest minute.

Nova Scotia

The current edition is dated 1978 and consequently is completely bilingual and contains a glossary in addition to the usual introductory material. In content the listings are standard for the newer gazetteers with the 1:50,000 sheet being given for each place name in addition to its geographical location. The glossary is extensive which is fitting because of the wealth of generic terms that have developed along the Atlantic coast. Sunkers (a solid mass of stone rising from the sea bottom, close to or above the surface), chuckles (a group of rocks at the mouth of a cove where tidal currents become agitated), etc. are not found elsewhere in Canada. The volume consists of 477 pages of listings.

Prince Edward Island

The most recent edition of the gazetteer of Canada's smallest province was published in 1973 to commemorate the 100th anniversary of the Island's entry into confederation. It was the first of the completely bilingual volumes in that all the introductory material is in both languages and all the descriptive notes in the listing are given in both languages. At first glance the listing does not appear to be bilingual because where the name includes the generic, as in West Reef or Winter River, no translation is given; but the name York is followed by (P.O. name—nom du B. de P.). As was mentioned in an earlier chapter, Prince Edward Island has its own peculiar designation of a rural administrative area that would be called a township in Ontario. This is the PEI lot, and in the gazetteer the lot and county is given for each place name. The 1:50,000 map on which the name falls is listed. The listing is 38 pages long.

Quebec

This is the only volume of the *Gazetteer of Canada* which is published by a provincial government rather than the federal authorities. In format and content it is much the same as the other gazetteers except that it is completely in French. This and the title *Répertoire Géographique du Québec* indicates Quebec's concern about the protection of the French language. In the days when the Quebec hinterland was being opened to colonization, the control of the administration of settlement was in British hands, and consequently an undue proportion of English place-names was installed. The effect was to misrepresent, in a way, the culture of the country. (To a certain extent the result of this mis-match of language of names and language of inhabitants can be seen in the New Brunswick volume where a casual glance through the pages of names would lead the reader to presume that almost all of the people of that province were of English origin whereas in fact about half of them are of French origin.) In any event the government of Quebec undertook a widespread gallicization of the provincial geographical names, and to establish early acceptance published the present gazetteer. It must be pointed out that this action was quite legal, as the provincial names authority has the final say on all names in the province.

The listings require 701 pages, and due to extensive cross referencing, a person familiar with the old English name or English spelling should have little trouble using the '*Répertoire*'. A typical cross reference is: Shickshock–Monts–voir Chick-Chocs 48°55′66°00′. The text is obtainable from the Ministère des Terres et Forêts du Québec.

Saskatchewan

This volume, published in 1969, has, like others published between 1967 and 1972, the introductory material in both languages but the listings in English only. The introductory material is brief, and no glossary is included. The listings include under the location column, the DLS township designation. The geographic location, like all other volumes except British Columbia, is to the nearest minute of latitude and longitude. The listings are contained in 173 pages.

Yukon Territory

This volume, published in 1976, is typical of the new style (since 1973) in having all material bilingual and extensive explanatory data including a glossary. The glossary is somewhat disappointing in that few of the colourful terms of the gold-rush days appear to have entered the English language. The term 'pup' used to describe a small intermittent tributary to a creek (as in Lombard Pup) seems to be restricted to the Yukon, but all other terms listed are in common use throughout Canada. The listings require 55 pages.

Appendix 1. Some Significant Dates in the Evolution of Canadian Mapping

1497 John Cabot's landfall in eastern Canada. His map has not survived.

1500 Juan de la Cosa published a world map embodying the cartographic record of John Cabot, the oldest map to show a part of Canada.

1535 First map of the Gulf of St Lawrence by Jean Rotz incorporating the results of Cartier's first voyage.

1537 Harleian map, the first to show the St Lawrence region and to use the name Canada.

1592 The terrestrial globe of Emery Molyneux includes the results of the 1585 explorations of John Davis.

1604 The first of Champlain's large-scale maps of Atlantic ports.

1607 Champlain's map delineating the coastline of Acadia.

1612 Gerritsz chart of Hudson Bay and Strait included Hudson's discoveries of 1610–11.

1613 Champlain makes first genuine attempt to lay down latitudes and longitudes on his map of the Atlantic coast.

1616 Champlain's map first to show any part of the Great Lakes based on European explorations.

1632 Champlain's last published map.

1635 Foxe's map shows outline of bay charted by William Baffin in 1616.

1642 Jean Bourdon continued Champlain's work with the publication of his 'Carte depuis Kebec jusque au Cap du Torments'.

1671 Jean Baptiste Louis Franquelin began producing maps of New France.

1709 Gideon de Catalogne's cadastral maps of the seigneuries along the St Lawrence River.

1731 The first known depiction of the southeastern part of the Canadian Prairies drawn by the Indian Ochagach for La Vérendrye.

1755 Dr John Mitchell publishes his 'Map of the British and French dominions in North America' used by the negotiators of the peace treaty between Great Britain and the United States to delimit the extent of British North America east of Lake of the Woods.

1758 Samuel Holland and James Cook work together to chart portions of the St Lawrence estuary.

1769 First scientific observations for latitude and longitude made in western Canada at Fort Prince of Wales (now Churchill) in preparation for observations of the transit of Venus.

1772 Samuel Hearne's 'Map of Part of the Inland Country to the Northwest of Prince of Wales Fort, Hudson's Bay' resulting from his expedition to the Arctic coast.

1778 Philip Turnor's map of the Canadian West, the first to be prepared from instrumental surveys, in that longitude readings were taken in addition to latitudes.

1783 Land survey in what is now Ontario commenced.

1785 Peter Pond's map discloses a potential canoe route into the Athabasca and Mackenzie river region.

1789 Alexander Mackenzie reaches the Arctic Ocean.

1793 Alexander Mackenzie reaches the Pacific Ocean overland and George Vancouver begins charting Canada's west coast.

1813–14 David Thompson's map of western Canada.

1842 The Geological Survey established and topographical mapping commenced.

1871 The Dominion Land Survey of the Prairies was commenced. Plans of survey were drawn at 1:31,680 (two inches to the mile).

1873 The Surveys Branch established within the Department of the Interior.

1883 The Surveys Branch renamed the Technical Branch.

1884 The Department of the Interior published a six miles to the inch map of the surveyed area of the Prairies compiled from the 1:31,680 DLS plans.

1886 Photo-topographical surveys commenced in the Rocky Mountains by the Technical Branch.

1886 The Department of Marine and Fisheries issued the first Canadian hydrographic chart.

1890 The Technical Branch, Department of the Interior, renamed the Topographical Surveys Branch.

1891 The first sheet of the three-mile series published.

1897 Geographic Board of Canada established primarily to regulate application of geographical names.

1903 Major E. H. Hills of the British Army was invited to examine and report upon the need in Canada for better and more extensive topographic mapping.

1903 The Chief Geographer's series at 1:250,000 and 1:500,000 introduced by the Department of the Interior.

1903 The Department of Militia and Defence established Canada's military mapping unit, titled the Mapping Branch of the Intelligence Department. Survey operations started in 1904.

1904 The Hydrographic Survey Branch established within the Department of Marine and Fisheries.

1906 First edition of *Atlas of Canada* published.

1906 The Mapping Branch, Department of Militia and Defence, renamed the Survey Division.
1906 The first two sheets of the one-inch series published.
1908 The Geological Survey grouped its topographers into the Topographical Survey Division.
1909 The Geodetic Survey of Canada officially established within the Department of the Interior.
1911 The first engraved Canadian coastal chart published.
1916 Military training maps at 1:20,000 and 1:31,680 introduced by Militia and Defence.
1922 The Board on Topographical Surveys and Maps formed to attempt coordination between the three topographic mapping agencies.
1922 The Topographical Surveys Branch, Department of the Interior, renamed the Topographical Survey of Canada.
1923 The first version of the National Topographic System was introduced by the Board on Topographical Surveys and Maps.
1924 The Survey Division, Militia and Defence renamed the Geographical Section of the General Staff (GSGS).
1925 The Board on Topographical Surveys and Maps replaced by the Board of Topographical and Aerial Surveys and Maps.
1926 The first sheet of the four-mile series was published.
1926 Canada's first aeronautical chart published.
1927 The second version of the National Topographic System was introduced.
1928 The Canadian Hydrographic Survey renamed the Canadian Hydrographic Service.
1929 The first eight-mile sheet (1:506,880) of the NTS was published.
1930 Arctic air route charting began.
1932 The 1:25,000 scale adopted for large-scale military mapping.
1933 The Topographical Survey of Canada renamed the Topographical and Air Survey Bureau.
1933 The Board of Topographical and Aerial Surveys and Maps changed to interdepartmental Committee on Air Surveys and Base Maps.
1936 The Department of the Interior disbanded; survey work taken over by the Department of Mines and Resources within its Bureau of Geology and Topography. A second mapping agency formed within the Department, titled Hydrographic and Map Service. Headed by the Surveyor-General it was responsible for hydrographic work, legal surveys of federal lands, aeronautical charting, map reproduction and sales.
1944 The NTS eight-mile series completed, though many northern sheets contained blank areas.
1946 The Geographical Section of the General Staff renamed the Army Survey Establishment.
1947 Canada's first serious long-range mapping programme approved by the Cabinet Defence Committee.
1948 Geographic Board of Canada becomes Canadian Board on Geographical Names.
1949 The NTS 1:253,440 scale changed to 1:250,000; the Universal Transverse Mercator projection adopted for all federal topographic mapping at scales of 1:250,000 and larger.

1950 The NTS 1:63,360 scale was changed to 1:50,000.
1950 The Department of Mines and Resources reorganized as the Department of Mines and Technical Surveys. Within this Department the Surveys and Mapping Branch was formed to include the Geodetic Survey, Legal Surveys and Aeronautical Charts, the Hydrographic Service, Topographical Survey and the Map Compilation and Reproduction Division.
1953 The 1:25,000 series brought into the National Topographic System.
1959 The Army Survey Establishment commenced work on the Military Town Plan series.
1960 *Atlas of Canada* first published in French (French version of the third edition of the national atlas).
1960 The marginal information on all federal topographic maps made bilingual.
1961 Canadian Board on Geographical Names reorganized as the Canadian Permanent Committee on Geographical Names.
1964 Publication of separate civilian and military versions of topographic maps discontinued. All copies of 1:25,000, 1:50,000 and 1:250,000 would now carry the UTM grid.
1966 The Army Survey Establishment renamed the Mapping and Charting Establishment.
1966 The Department of Mines and Technical Surveys renamed the Department of Energy, Mines and Resources. The Canadian Hydrographic Service left the Surveys and Mapping Branch for the newly created Department of the Environment.
1968 Geodetic Survey absorbs the field survey component of Topographical Survey.
1970 Final sheet of the 1:250,000 series published.
1977 All federal topographic maps to be completely bilingual as they are produced after this date.

Appendix 2. Minimum Dimensions of Features for Inclusion on Topographic Maps

The term diameter is used loosely in this table to mean the average width of a feature that is roughly round in shape. When two equal dimensions (e.g., 60 × 60) are given, a feature of equivalent area, though not square will be shown. All dimensions are in metres.

Feature	1:25,000	1:50,000	1:250,000
Cutting or embankment (on roads or railroads)			
Length	50	100	Not shown
Ditch			
Length	300	600	3000
Width of double line ditch	10	20	100
Esker			
Length	200	400	2000
Filtration bed			
Width and length	60 × 80	120 × 160	600 × 800
Flooded land			
Diameter	100	200	1000
Foreshore flats			
Width	25	50	250
Glacier			
Width and length	50 × 150	100 × 300	500 × 1500
Hedgerow			
Length	300	Not shown	Not shown
Inundated land			
Width and length	50 × 100	100 × 200	500 × 1000
Island			
Diameter	8	15	75
(Islands smaller than minimum size are shown by the rock or reef symbol.)			
Lake or pond			
Diameter	10	20	300
Marsh or swamp			
Diameter	60	120	600

Feature	1:25,000	1:50,000	1:250,000
Mine waste			
Diameter	60	120	600
Moraine			
Width and length	50 × 100	100 × 200	500 × 1000
Orchard			
General dimensions	60 × 60	120 × 120	Not shown
Peat cutting			
General dimensions	300 × 300	600 × 600	Not shown
Quarry			
Diameter	60	120	600
Reservoir			
Width and length	15 × 30	30 × 60	150 × 300
(If reservoir is an artificial lake, the lake minimums are used.)			
Sand beaches			
Width	25	50	250
Sandy areas			
Diameter	60	120	600
Sewage disposal bed			
Width and length	60 × 80	120 × 160	Not shown
Ski lift			
Length of permanent installation	All shown	All shown	If over 1500
Slag heap			
Diameter	60	120	600
Streams			
Length	300	600	3000
Width of double line stream	10	20	100
Tree nursery			
General dimensions	60 × 60	120 × 120	Not shown
Tundra polygons			
Width and length	200 × 300	400 × 600	2000 × 3000
Wooded areas or cleared areas in woods			
Diameter	60	120	600
Vineyards			
General dimensions	60 × 60	120 × 120	Not shown

These minimum dimensions are at times treated with cartographic discretion. For example on Sheet 34 G (Lake Minto) where there are over 10,000 lakes and ponds and an equal number of islands, some of these features may not be shown even though they are larger than the minimum size. On the other hand, in the dry-lands of southern Alberta a small pond or water hole would be a landmark feature and would be shown even if slightly below the minimum size. On the map, features at or near the minimum size are depicted at a slightly enlarged scale so they can be seen.

Appendix 3. Contour Intervals on Federal Topographic Maps and Type Specifications for Populated Places

Terrain type	1:25,000	1:50,000	1:250,000
Imperial units—contour intervals in feet			
Very flat to normal	10	25	100
Hilly	25	50	200
Mountainous	Not used	100	500
Metric units—contour intervals in metres			
Very flat	2.5	5	20
Normal	5	10	50
Hilly	10	20	100
Mountainous	20	40	200

Every fifth contour is an index contour.

Type Specifications for Populated Places (all Century Schoolbook)

	1:25,000 and 1:50,000	1:250,000
City over 100,000	18 pt. caps Roman	14 pt. u/l Roman
City 25,000 to 100,000	14 pt. caps Roman	12 pt. u/l Roman
Town 10,000 to 25,000	12 pt. caps Roman	10 pt. u/l Roman
Town 2000 to 10,000	12 pt. u/l Roman	9 pt. u/l Roman
Villages 500 to 2000	10 pt. u/l Roman	8 pt. u/l Roman
Villages 100 to 500	8 pt. u/l Roman	7 pt. u/l Roman
Villages or settlements under 100	8 pt. u/l Italic	7 pt. u/l Roman

Caps = capital letters
U/l = upper and lower case

Appendix 4. Map Coverage of the Provinces and Territories of Canada

Province or territory	Land	Freshwater	Total	Percentage of total area	1:50,000 Mapping: Quadrangles—total coverage	1:50,000 Mapping: Published by 1 April /79	1:50,000 Mapping: Remaining
	(square miles)	(square miles)	(square miles)				
Newfoundland	143,045	13,140	156,185	4.1			
Island of Newfoundland	41,164	2,195	43,359	1.1	128	128	0
Labrador	101,881	10,945	112,826	3.0	416	408	8
Prince Edward Island	2,184	—	2,184	0.1	13	13	0
Nova Scotia	20,402	1,023	21,425	0.6	87	87	0
New Brunswick	27,835	519	28,354	0.7	85	85	0
Quebec	523,860	71,000	594,860	15.4	1,752	1,652	100
Ontario	344,092	68,490	412,582	10.7	1,095	773	322
Manitoba	211,775	39,225	251,000	6.5	749	414	335
Saskatchewan	220,182	31,518	251,700	6.5	723	697	26
Alberta	248,800	6,485	255,285	6.6	749	643	106
British Columbia	359,279	6,976	366,255	9.5	1,162	976	186
Yukon Territory	205,346	1,730	207,076	5.4	685	443	242
Northwest Territories	1,253,438	51,465	1,304,903	33.9	5,506	1,150	4,356
Franklin	541,753	7,500	549,253	14.3			
Keewatin	218,460	9,700	228,160	5.9			
Mackenzie	493,225	34,265	527,490	13.7			
Total for Canada	3,560,238	291,571	3,851,809	100.0	13,150	7,469	5,681

Appendix 5. Federal and Provincial Policies on Bilingualization and Metrication

1. *Federal policies*

   (a) *Bilingualization*

   In 1960 the first step was taken toward producing bilingual topographic maps. A bilingual design for all marginal data was produced which has been used for all federal topographic maps since that year. On the face of the map, the generic part of each place name (i.e. lake, river, lac, fleuve, etc.) was written in the way it was most commonly used in the area, in French or English. Some large features were named in both French and English, particularly if they ran between English and French speaking communities, such as the Ottawa River—Riviere Outouais. All labels on the map, descriptive words such as tower, trailer camp, shopping centre, etc., remained in English.

   This situation persisted until 1977 when the decision was made to make the complete topographic map bilingual. The labelling was the only problem, and it turned out to be not a major one. To avoid labelling in both languages, the most common labels are translated in a glossary in the margin, so that only one term, French in Quebec, English in the rest of Canada, need be used on the face of the map. The unusual terms, for example 'Boy Scout Camp' or 'Mink Farm', will be labelled in both languages. It will of course be many years before existing sheets are converted to the bilingual format, but all sheets coming up for revision will be bilingal in the new editions.

   For maps smaller in scale than 1:1,000,000, two editions will be published, one completely in French, the other in English.

   (b) *Metrication*

   Metrication in the scale of federal topographic maps has been a fact since the beginning of the 1:25,000, 1:50,000 and 1:250,000 series, but until 1975 all contouring was done in even intervals of feet above sea-level. The destination notes in the margin, giving the distance to the next town or city, were in miles. Since 1975 the contouring of all new maps has been changed to metres. Distances are now given in kilometres, spot heights in metres. As with the move to bilingualism, the complete conversion to the new style will take many years. Appendix 3 gives the contour intervals used on Canadian topographic maps at various scales.

2. *Provincial Policies*

(a) *Metrication*

The provincial mapping authorities were somewhat slower than the federal government in adopting metric contours and metric scales, but by 1978 all were producing some metric maps. Resource mapping, such as the large-scale topographic maps produced by Hydro Quebec for the immediate areas around power dam construction sites, are very closely attuned to the practice in the construction industry which must make estimates of rock cuts, etc. in the familiar imperial units. The conversion of such maps tends to lag behind. All the provincial road maps, except that of Prince Edward Island, are in metric form. Metric topographic and thematic maps will come in time.

(b) *Bilingualization*

Only one of Canada's provinces is officially bilingual—New Brunswick. Quebec publishes all official maps in French; the other provinces in English, though a few bilingual maps are found. A trend to more bilingual maps is difficult to perceive at the present time.

Appendix 6. Map Scales and Equivalents

For assistance in determining map scales and their equivalent in miles or feet per inch

Fractional scale	Miles per inch	Inches per mile	Feet per inch
1:500	0.008	126.72	41.67
1:600	0.009	105.60	50.00
1:1,000	0.016	53.36	83.33
1:1,200	0.019	52.80	100.00
1:1,500	0.024	42.24	125.00
1:2,000	0.032	31.68	166.67
1:2,400	0.038	26.40	200.00
1:2,500	0.039	25.34	208.33
1:3,000	0.047	21.12	250.00
1:3,600	0.057	17.60	300.00
1:4,000	0.063	15.84	333.33
1:4,800	0.076	13.20	400.00
1:5,000	0.079	12.67	416.67
1:6,000	0.095	10.56	500.00
1:7,000	0.110	9.05	583.33
1:7,200	0.114	8.80	600.00
1:7,920	0.125	8.00	660.00
1:8,000	0.126	7.92	666.67
1:8,400	0.133	7.54	700.00
1:9,000	0.142	7.04	750.00
1:9,600	0.152	6.60	800.00
1:10,000	0.158	6.34	833.33
1:10,800	0.170	5.87	900.00
1:12,000	0.189	5.28	1,000.00
1:13,200	0.208	4.80	1,100.00
1:14,400	0.227	4.40	1,200.00
1:15,000	0.237	4.22	1,250.00
1:15,600	0.246	4.06	1,300.00
1:15,840	0.250	4.00	1,320.00
1:16,000	0.253	3.96	1,333.33

Fractional scale	Miles per inch	Inches per mile	Feet per inch
1:16,800	0.265	3.77	1,400.00
1:18,000	0.284	3.52	1,500.00
1:19,200	0.303	3.30	1,600.00
1:20,000	0.316	3.17	1,666.67
1:20,400	0.322	3.11	1,700.00
1:21,120	0.333	3.00	1,760.00
1:21,600	0.341	2.93	1,800.00
1:22,800	0.360	2.78	1,900.00
1:24,000	0.379	2.64	2,000.00
1:25,000	0.395	2.53	2,083.33
1:31,680	0.500	2.00	2,640.00
1:48,000	0.758	1.32	4,000.00
1:50,000	0.789	1.27	4,166.67
1:62,500	0.986	1.01	5,208.33
1:63,360	1.000	1.00	5,280.00
1:75,000	1.184	0.85	6,250.00
1:96,000	1.515	0.66	8,000.00
1:100,000	1.578	0.63	8,333.33
1:125,000	1.973	0.51	10,416.67
1:126,720	2.000	0.50	10,560.00
1:200,000	3.157	0.32	16,666.67
1:250,000	3.946	0.25	20,833.33
1:253,440	4.000	0.25	21,120.00
1:380,160	6.000	0.17	31,680.00
1:400,000	6.313	0.16	33,333.33
1:500,000	7.891	0.13	41,666.67
1:666,667	10.521	0.09	55,555.55
1:750,000	11.837	0.08	62,500.00
1:760,320	12.000	0.08	63,360.00
1:1,000,000	15.783	0.06	83,333.33
1:1,013,760	16.000	0.06	84,480.00
1:1,267,200	20.000	0.05	105,600.00
1:1,500,000	23.674	0.04	125,000.00
1:1,520,640	24.000	0.04	126,720.00
1:1,680,000	26.515	0.04	140,000.00
1:2,000,000	31.565	0.03	166,666.67
1:2,217,600	35.000	0.03	184,800.00
1:2,500,000	39.457	0.03	208,333.33
1:3,000,000	47.348	0.02	250,000.00
1:3,500,000	55.240	0.02	291,666.67
1:4,000,000	63.131	0.02	333,333.33
1:4,055,040	64.000	0.02	337,920.00
1:4,500,000	71.023	0.01	375,000.00
1:4,561,920	72.000	0.01	380,160.00
1:4,752,000	75.000	0.01	396,000.00
1:5,000,000	78.914	0.01	416,666.67
1:5,068,800	80.000	0.01	422,400.00
1:6,000,000	94.697	0.01	500,000.00
1:7,000,000	110.479	0.01	583,333.33
1:7,500,000	118.371	0.01	625,000.00
1:7,920,000	125.000	0.01	660,000.00
1:8,000,000	126.262	0.01	666,666.67
1:9,000,000	142.045	0.01	750,000.00
1:9,187,200	145.000	0.01	765,600.00
1:10,000,000	157.828	0.01	833,333.33
1:11,000,000	173.611	0.01	916,666.67
1:11,088,000	175.000	0.01	924,000.00
1:12,000,000	189.393	0.01	1,000,000.00
1:12,500,000	197.280	0.01	1,041,666.00

Fractional scale	Miles per inch	Inches per mile	Feet per inch
1:13,000,000	205.176		1,083,333.33
1:14,000,000	220.959		1,116,666.67
1:15,000,000	236.742		1,250,000.00
1:16,000,000	252.525		1,333,333.33
1:17,000,000	268.308		1,416,666.67
1:18,000,000	284.090		1,500,000.00
1:19,000,000	299.873		1,583,333.33
1:20,000,000	315.656		1,666,666.67
1:21,000,000	331.439		1,750,000.00
1:22,000,000	347.222		1,833,333.33
1:23,000,000	363.005		1,916,666.67
1:24,000,000	378.787		2,000,000.00
1:25,000,000	394.570		2,083,333.33

Appendix 7. General Maps

Map No.	Title	Scale	Colours	Size	Year
MCR 18	Canada	1″–250 miles	7	17″ × 14″	1973
MCR 19	Canada (Fr.)	1″–250 miles	7	17″ × 14″	1973
MCR 22 (without names)	Canada (sans désignation)	1″–250 miles	1	17″ × 14″	1962
MCR 64	Canada	1″–250 miles	2	17″ × 14″	1970
MCR 15	Canada	1″–140 miles	7	30″ × 27″	1970
MCR 16	Canada (Fr.)	1″–140 miles	7	30″ × 27″	1970
MCR 78 (without names)	Canada (sans désignation)	1″–105.8 miles	1	42″ × 31″	
MCR 10	Canada	1″–100 miles	7	40″ × 36″	1970
MCR 11	Canada (Fr.)	1″–100 miles	7	40″ × 36″	1970
MCR 8	Canada	1″–64 miles	7	58½″ × 62″	1966
MCR 8F	Canada	1″–64 miles	7	58½″ × 62″	1966
MCR 5 (6 sheets/feuilles)	Canada	1:2,000,000	10	42″ × 60″ each chacune	1971
MCR 82 (6 sheets/feuilles)	Canada (Fr.)	1:2,000,000	10	42″ × 60″ each chacune	1971
MCR 31	North America	1:10,000,000	8	32″ × 40″	1971
MCR 31F	Amérique du Nord	1:10,000,000	8	32″ × 40″	1971
MCR 36	Northwest Territories and Yukon Territory (shaded) (ombrée)	1:4,000,000	8	42″ × 52″	1974
MCR 35	Northwestern Canada (transportation/voies de transport)	1″–50 miles	3	25″ × 31½″	1974

Map No.	Title	Scale	Colours	Size	Year
MCR 47	Yukon Territory	1:2,000,000	5	22″ × 26″	1976
MCR 3	British Columbia	1:2,000,000	5	31″ × 42″	1973
MCR 27	Prairie Provinces	1:2,000,000	5	31″ × 42″	1973
MCR 39	Ontario	1:2,000,000	5	36″ × 40″	1973
MCR 42	Québec	1:2,000,000	5	38″ × 42″	1973
MCR 42F	Québec (Fr.)	1:2,000,000	5	38″ × 42″	1973
MCR 77	Atlantic Provinces	1:2,000,000	5	31″ × 42″	1973
MCR 40 (outline) (tracé)	Ontario South	1:1,000,000	1	34″ × 24″	1965
MCR 29	New Brunswick	1:500,000	7	38″ × 36″	1972
MCR 30	Newfoundland	1:500,000	5	36″ × 43″	1975
MCR 41	Prince Edward Island	1:250,000	6	25½″ × 36″	1974
MCR 37	Nova Scotia and Prince Edward Island	1:500,000	7	38″ × 50″	1974
MCR 38	Maritime Provinces	1″–10 miles	6	37″ × 51″	1968
MCR 26 (2 sheets/feuilles)	Manitoba	1″–12 miles	4	N42″ × 36¼″ S46½″ × 42″	1964
MCR 45 (2 sheets/feuilles)	Saskatchewan	1″–12 miles	5	N38″ × 38¼″ S38″ × 46¼″	1963
MCR 83 (2 sheets/feuilles)	Alberta	1:750,000	9	42″ × 36″	1972
MCR 25	Yukon Territory	1:1,000,000	8	40″ × 50″	1972
MCR 2	Arctic North America	1″–200 miles	2	20″ × 12″	1964
MCR 46	The World	1:35,000,000	Multi-	48″ × 36″	1974
MCR 46F	Le Monde	1:35,000,000	Multi-	48″ × 36″	1974
MCR 34	The Northern Hemisphere	1″–300 miles	3	40″ × 36″	1965
MCR 34F	Hémisphère Nord	1″–300 miles	3	40″ × 36″	1965
MCR 7	Centennial Range	1:125,000	4	30″ × 34″	1967

All of the above maps are available through the Canada Map Office

Appendix 8. The Availability of Official Canadian Maps, Charts, Atlases and Gazetteers

Maps of the Federal Government

All current topographic maps produced by the federal government are available from:

The Canada Map Office,
Surveys and Mapping Branch,
Department of Energy, Mines and Resources,
Ottawa, Ontario,
K1A 0E9.

This office also handles photo maps, aeronautical charts, general maps of Canada and the provinces at small scale, maps of national parks, and Canada land inventory maps—soil capability for agriculture, generalized land use, land capability for forestry, land capability for recreation, land capability for wildlife–ungulates, land capability for wildlife–waterfowl, as well as series produced by other departments of the federal government. Map indexes are provided free on request.

Other important federal map producers are listed below with the types of maps they have for sale. The availability of sales literature (catalogues, indexes, etc.) is indicated in brackets.

Agency	*Products*
Publications and Information Office, Geological Survey of Canada, Department of Energy, Mines and Resources, 601 Booth Street, Ottawa, Ontario, K1A 0E8.	Geological maps Geophysical maps (Catalogue of publications)

Hydrographic Chart Distribution Office, Marine Sciences Branch, Fisheries and Environment Canada, Ottawa, Ontario, K1G 3H6.	Hydrographic charts Bathymetric charts Marine geological charts Marine natural resource maps (Chart indexes)
The Lands Directorate, Fisheries and Environment Canada, Ottawa, Ontario, K1A 0E7.	Land Use Information series 1:250,000 Land capability maps (Brochure)
Publications and Information Office, Agriculture Canada, Ottawa, Ontario, K1A 0C7.	Soils maps Agricultural land use maps (List of publications)

Provincial Maps

All provinces have several map-producing agencies. The more important of these are listed, together with the type of maps they produce. The availability of sales literature is indicated in brackets.

Alberta

Agency	*Products*
Surveys and Mapping Branch, Alberta Transportation, Highway Building, Edmonton, Alberta, T5K 2B8.	Provincial 1:250,000 series Alberta oil sands topographic maps Cadastral plans Highway development maps City maps (Maps and plans catalogue)
Geological Division, Alberta Research, Edmonton, Alberta, T6G 2C2.	Geological maps Geophysical maps (List of publications)
Alberta Map and Air Photo Distribution Centre, Alberta Energy and Natural Resources, 9945–108th Street, Edmonton, Alberta, T5K 2G6.	Topographic maps Planimetric maps Forest cover maps Township plans Separate sheets from the *Atlas of Alberta* (Map catalogue)

Energy Resource Conservation Board, 603 Sixth Avenue, SW, Calgary, Alberta, T2P 0T4.	Oil and gas development maps Electrical power development maps Coal production planning maps (Publications, Maps and Services catalogue)

British Columbia

Agency	*Products*
Map and Air Photo Sales Surveys and Mapping Branch, Ministry of the Environment, Victoria, BC, V8V 1X5.	Topographical maps Planimetric maps Photomaps Cadastral plans (Map and air photo catalogue)
Publications Office, Department of Mines and Petroleum Resources, Victoria, BC, V8V 1X4.	Geological maps Mineral inventory maps (Map catalogue and map indexes)
Resources and Analysis Branch, Ministry of the Environment, Victoria, BC, V8V 1X4. (Note: This agency was formerly known as the Environment and Land Use Secretariat).	Soils maps Forestry maps Land capability maps (Catalogue of maps and publications)

Manitoba

Agency	*Products*
Map Sales Office, Surveys and Mapping Branch, Department of Renewable Resources and Transportation Services, 1007 Century Street, Winnipeg, Manitoba, R3H 0W4.	Topographic maps of northern settlements Cadastral plans and township plans of southern Manitoba Forest inventory maps Lake depth charts (Map catalogue)
Information and Publication Office, Mineral Resources Branch, Department of Mines, Resources and Environmental Management, Winnipeg, Manitoba, R3C 0P8.	Mineral claim maps Geological maps (Index of publications)

Maritime Provinces (Nova Scotia, New Brunswick and Prince Edward Island)

Agency	*Products*
Surveys and Mapping Division, Land Registration and Information Service, 120 Water Street East, Summerside, PEI, C1N 1A9.	Large-scale topographic maps Resource maps (Resource and large-scale map index)
Maritime Resources Management Service, PO Box 310, Amherst, NS, B4H 3Z5.	Thematic maps (Map catalogue)
Director Technical Services Branch, Department of Agriculture, Box 1600, Charlottetown, PEI, C1A 7N8.	Soil maps of Prince Edward Island (List of publications)
Mineral Resources Branch, Department of Natural Resources, Fredericton, NB. E3B 5H1.	Geological maps of New Brunswick (List of publications)
Lands Branch, Department of Natural Resources, Fredericton, NB, E3B 5H1.	Forestry maps of New Brunswick (List of publications)

Newfoundland and Labrador

Agency	*Products*
Department of Mines, Energy and Mineral Development, 95 Bonaventure Avenue, St. John's, Newfoundland, A1C 5T7.	Geological maps (List of publications)
Department of Forestry and Agriculture, Forest Inventory Division, PO Box 235, St. John's, Newfoundland, A1C 5J2.	Forestry maps (List of publications)

Department of Municipal Affairs and Housing, Confederation Building, Box 4750, St. John's, Newfoundland, A1C 5T7.	Plans of cities, towns and villages (List of publications)

Ontario

Agency	*Products*
Public Service Centre, Ministry of Natural Resources, Administrative Services Branch, Queen's Park, Toronto, Ontario, M7A 1W3.	Ontario base maps at 1:20,000, 1:10,000, 1:2000 Planimetric maps Geological maps Resource planning maps (Map catalogue and map indexes)
Survey Records Office, Ministry of Natural Resources, Surveys and Mapping Branch, Queen's Park, Toronto, Ontario, M7A 1W3.	Township plans Park plans Game reserves (List of publications)
Cartography Section, Ministry of Transportation and Communications Downsview, Ontario, M3M 1J8,	1:250,000 county maps Road maps Highway development maps (List of publications)
Ministry of Treasury, Economics and Intergovernmental Affairs, Queen's Park, Toronto, Ontario, M7A 1Y7.	Regional development maps Land use maps (List of publications)
Hydrology Branch, Ministry of the Environment, 135 St. Clair Avenue West, Toronto, Ontario, M4V 1P5	Hydrological maps Ground water maps Hydro-geological maps (List of reports and maps)

Quebec

Agency	*Products*
Ministère des Terres et Forêts, Service de la Cartographie, 1995 Ouest, boul. Charest, Québec, PQ, G1N 4H9.	Large-scale topographic maps 1:200,000 cadastral maps Forestry maps Hydrological maps Planimetric maps City maps County maps Cadastral maps (Catalogue – répertoire)
Ministère des Richesses Naturelles, Distribution et documentation, 1620 boul. de l'Entente, Québec, PQ, G1S 4N6.	Geological maps (List of publications)
Ministère d'Agriculture, 200–A Chemin Ste-Foy, Québec, PQ, G1R 4X6.	Agricultural land use, soil maps (List of publications)
Office de planification et de development du Québec, 1050 St-Augustin, Québec, PQ, G1R 4Z5.	Thematic maps (List of publications)
Ministère des Transports, Division de la Cartographie, Service de la Géographie, Québec, PQ, G1R 4Y8.	1:125,000 road maps Provincial road maps Snowmobile maps (Sentiers de Motoneige) 1 inch = 3 miles (List of publications)

Saskatchewan

Agency	*Products*
Publications Office, Department of Mineral Resources, PO Box 5114, Regina, Saskatchewan, S4P 3P5.	Provincial geological and geophysical maps (Map catalogue)

Lands and Surveys Branch, Department of Tourism and Renewable Resources, 1840 Lorne Street, Regina, Saskatchewan, S4P 2L7.	Forest inventory maps Tourist maps (List of publications)
Central Surveys and Mapping Agency, Department of Highways and Transportation, Regina, Saskatchewan, S4S 0B1.	Large-scale topographic maps and plans (List of publications)

Note: Most maps of the Yukon and the Northwest Territories are produced by federal government mapping agencies.

National and Provincial Atlases

The fourth edition of the *Atlas of Canada* is available through the Canada Map Office. Copies of the preceding three editions are out of print, but are seen from time to time in used book stores.

Provincial atlases are in a number of cases out of print. Those available at the time of writing are:

Atlas of Alberta	University of Alberta Press, Edmonton, Alberta, T6G 2E8.
Atlas of British Columbia	University of British Columbia Press, Vancouver, BC, V6T 1W5
Economic Atlas of Ontario	University of Toronto Press, Toronto, Ontario, M5S 1A1.
Atlas of Saskatchewan	University of Saskatchewan, Saskatoon, Saskatchewan, S7N 0W0.

Historic Maps and Superseded Editions

The foregoing listings show the source of current editions of Canadian maps, but none of the agencies mentioned make a practice of selling superseded editions of their maps. Most map producers keep a collection of their early editions and generally arrangements can be made to examine these on the premises. Almost all of the maps produced by federal agencies, including most back editions, are found in the National Map Collection of the Public Archives of Canada. This remarkable collection also possesses most of the significant historic maps mentioned in Chapter 1, and a good collection of the provincial mapping described in Chapter 13. Photocopies of these maps will be provided by the Public Archives staff at cost.

Provincial archives and university map colletions are also places where old maps may be found and examined. Some of these institutions also provide photocopying services. Provincial archives are, understandably, the best place to look for old editions of provincial mapping.

Canada is fortunate in having a large number of private map dealers, but most of these carry only the current editions. However, many of the larger cities have one or two antique map dealers who carry both very old and expensive maps and some that are not quite so old such as the three-mile and the Chief Geographer's series.

Replicas of historic Canadian maps are available at modest cost from two sources. The following are available from the Canada Map Office, Ottawa, Ontario:

America Septentrionalis (Jan Jansson) 1639
Le Canada, ou Nouvelle France (Nicolas Sanson) 1656
Amerique Septentrionale (Alexis Hubert Jaillot) c 1696
An Accurate Map of Canada (J. Hinton) c 1761
British North America (J. Arrowsmith) 1834
Map of Mounted Police Stations and Patrols for 1888

The following are available from the Association of Canadian Map Libraries, c/o National Map Collection, Public Archives of Canada, Ottawa, Ontario, K1A 0N3:

America (Arnoldo Di Arnoldi) 1600
The North Part of America (Henry Briggs) 1625
Le Canada Faict Par Le Sr. De Champlain ou Sont La Nouvelle France, La Nouvelle Angleterre, La Nouvelle Holande, La Nouvelle Suede, La Virginie (Pierre Du Val) 1653
Lac Superieur et Autres Lieux où Sont Les Missions Des Péres De La Compagnie de Jésus Comprises Sous Le Nom D'Outaouacs (Claude Dablon) 1673
A Map of the North Pole and the Parts Adjoining (Moses Pitt) 1680
A New Chart of the Coast of New England, Nova Scotia, New France, or Canada, with the Islands of Newfoundland, Cape Breton, St. John's, etc. (J-N. Bellin) 1746
A Map of the South Part Nova Scotia and It's Fishing Banks (Thomas Jefferys) 1750
Partie Occidentale Du Canada, Contenant Les Lacs Ontario, Huron, Erie, et Lac Superieur (J-N. Bellin) 1752
Nouvelle Carte Des Decouvertes Faites Par Des Vaisseaux Russes Aux Côtes Inconnues De L'Amerique Septentionale Avec Les Pais Adjacents 1754
A Map of the Island of St. John (Thomas Jefferys) 1775
Chart of the N.W. Coast of America and N.E. Coast of Asia Explored in the Years 1778 and 1779 (Capt. James Cook's voyage) 1784
Western Canada (Peter Pond) 1785
A New Map of the World, with the Latest Discoveries (Samuel Dunn) 1794
A Map of America . . . Exhibiting MacKenzie's Track (Alexander MacKenzie) 1801

Plan De La Cataracte De Niagara et De L'Isthme Qui Separe Les Lacs Erie et Ontario (P. F. Tardieu) 1805
A Map of the Province of Upper Canada, Describing All the New Settlements, Townships, etc. With the Countries Adjacent from Quebec to Lake Huron (David William Smyth) 1813
Toronto, Canada West (Waterlow and Sons Lith.) 1857
British Columbia (J. Conroy) 1862
Province of Manitoba and the Part of the District of Keewatin and Northwest Territory Showing the Townships and Settlements Drawn From the Latest Gov. Maps, Surveys and Reports (H. Beldon & Co.) 1879

This is a continuing series and additions are made to it at frequent intervals.

Map dealers almost always include current and old atlases in their stock. As atlases are more durable than maps, more of them have survived than maps of equivalent age. County atlases appear for sale from time to time and if one is patient the atlas of one's choice can usually be obtained. Knowledge of the value of these atlases has spread in recent years and prices have risen accordingly. However, there is a factor that keeps these prices within reason and this is the availability of most of them in reprint editions. The two most active publishers are listed below with the county atlases they have produced.

Cumming Atlas Reprints, Box 23, Stratford, Ontario, NSA 658

Eastern Townships and Southwest Quebec
Bruce
Carleton
Elgin
Essex and Kent
Grey
Haldimand
Halton
Huron
Lambton
Lanark and Renfrew
Lincoln and Welland
Middlesex
Muskoka and Parry Sound
Ontario
Oxford
Peel
Perth
Simcoe
Stormont, Dundas, Glengarry, Prescott and Russell
Waterloo and Wellington
Wentworth

Mika Publishing Co., Box 536, Belleville, Ontario, KN8 5B2

King's, Queen's and Prince County (PEI)
Pictou (NS)
Saint John and York (NB)
Frontenac, Lennox and Addington
Hastings and Prince Edward
Middlesex
Brant
Huron
Norfolk
Northumberland and Durham
Ontario
Oxford
Perth
York (Ont.)
Leeds and Grenville
Wellington
Stormont, Dundas and Glengarry

Gazetteers

All gazetteers except that of Quebec, can be purchased from the Publication Centre, Department of Supply and Services, Ottawa, Ontario, Canada, K1A 0S9. Each year a supplement to each volume is published recording all newly approved names and all name changes that have taken place since the most recent edition was published. These cumulative supplements are provided by the Canada Map Office, free of charge on request, to all purchasers of each gazetteer.

The Répertoire Géographique du Québec is obtainable from:

Ministère des Terres et Forêts,
Service de la Cartographie,
1995 Ouest, boul. Charest,
Québec, PQ,
G1N 4H9.

Aerial Photography

Aerial photographs can often be used to augment or amplify the information obtained from topographic maps. All of Canada has been photographed from the air. The cameras used in this work were all calibrated, and the resulting photography has all been inspected to ensure its dimensional accuracy. The flight lines which overlap by about 30% are usually north–south or east–west. Along the flight line the photographs overlap by about 60% so that steroscopic viewing is always possible.

Although photo-coverage of Canada is complete, each year certain areas are reflown, (a) to record the changes in the topography made by man or nature, and (b) to improve the quality of existing photography. There is, therefore, a choice of scales and dates for most parts of Canada.

Negatives for all photographs taken by the federal government are held by the Surveys and Mapping Branch. Prints from these negatives, contact or enlarged, may be obtained from the National Air Photo Library, 615 Booth Street, Ottawa, Ontario, K1A 0E9.

Notes and References

Chapter 1 An Outline of the Mapping of Canada

1. J. Spink and D. W. Moodie, 'Eskimo Maps from the Canadian Eastern Arctic', *Cartographica*, 5, 1972.
2. W. F. Ganong, 'Crucial Maps in the Early Cartography and Place-Nomenclature of the Atlantic Coast of Canada', University of Toronto Press, Toronto 1964, p. 11.
3. *Ibid.*, p. 221.
4. R. A. Skelton, *Explorers' Maps*, Routledge and Kegan Paul, London 1958, p. 123.
5. J. Warkentin and Richard I. Ruggles, *Historical Atlas of Manitoba*, The Historical and Scientific Society of Manitoba, Winnipeg 1970.
6. C. E. Heidenreich, 'Explorations and Mapping of Samuel de Champlain, 1603–1632', *Cartographica*, 17, 1976.
7. N. L. Nicholson, *The Boundaries of the Canadian Confederation*, Macmillan, Toronto 1979.
8. J. Warkentin and Richard I. Ruggles, *Historical Atlas of Manitoba*, The Historical and Scientific Society of Manitoba, Winnipeg 1970, p. 62.
9. Don W. Thomson, *Men and Meridians: The History of Surveying and Mapping in Canada*, Vol 1, Queen's Printer, Ottawa 1967, p. 105.
10. J. Warkentin and Richard I. Ruggles, *Historical Atlas of Manitoba*, The Historical and Scientific Society of Manitoba, Winnipeg 1970, p. 92.
11. Don, W. Thomson, *Men and Meridians: The History of Surveying and Mapping in Canada*, Vol 2, Queen's Printer, Ottawa 1968.
12. W. Kaye Lamb (ed.), *The Journals and Letters of Sir Alexander Mackenzie*, Macmillan, Toronto 1970.
13. Coolie Verner, 'The Arrowsmith Firm and the Cartography of Canada', *The (Canadian) Cartographer*, 8, 1971, pp. 1–7.
14. A. Taylor, *Geographical Discovery and Exploration in The Queen Elizabeth Islands*, Queen's Printer, Ottawa 1955.
15. Don. W. Thomson, *Men and Meridians: The History of Surveying and Mapping Canada*, Vol 2, Queen's Printer, Ottawa 1968.
16. Major E. H. Hills, *Report on the Survey of Canada*, War Office, London 1903.
17. J. K. Fraser, 'The Islands in Foxe Basin', *Geographica' Bulletin*, 4, 1953, pp. 1–31.

Chapter 2 Systematic Mapping before 1890

1. H. Maddick and J. Winearls, *County Maps*, Public Archives of Canada, Ottawa 1976.
2. Willis Chipman, 'A Plea for a Topographical Survey', *Annual Report of the Association of Ontario Land Surveyors*, 1893, pp. 87–94.

3. Maddick and Winearls, *op. cit.*
4. Data obtained from Betty May, *County Atlases of Canada*, Public Archives of Canada, Ottawa 1970.
5. Willoughby Vernor, *Rapid Field Sketching and Reconnaissance*, W. H. Allen, London 1889.
6. A description of this campaign is given in Major G. Dennison, *The Fenian Raid on Fort Erie*, Rolls and Adams, Toronto 1866.
7. Board on Topographical Surveys and Maps, Interior File 17884, 7 September 1920. (Paper written by Surveyor-General E. G. Deville).

Chapter 3 The Three-Mile Sectional Maps of the Canadian West

1. Complete details of the Dominion Land Survey system are given in: *Manual showing the System of Survey adopted for the Public Lands of Canada*, Secretary of State for Canada, 1st edition 1871; *Manual showing the System of Survey of the Dominion Lands*, Interior, 2nd edition 1881, 3rd edition 1883, 4th edition 1890; *Manual of Instructions for the Survey of Dominion Lands*, Interior, 5th edition 1903, 6th edition 1905, 7th edition 1910, 8th edition 1913, 9th edition 1918, 10th edition 1946.
2. *Manual showing the System of Survey of the Dominion Lands*, Interior, 2nd edition 1881.
3. By 1948 only two contoured four-mile sheets had been published for areas also covered by three-mile sheets. These were 62 H Winnipeg and 62 I Selkirk which were published in 1930 and 1929 respectively.

Chapter 4 The One Inch to One Mile Series

1. Major E. H. Hills, *Report on the Survey of Canada*, War Office, London 1903.
2. J. Sutherland Brown, 'The Military Survey of Canada', *Canadian Defence Quarterly*, 2, 1924.
3. Board on Topographical Surveys and Maps, Interior File 17884, 7 September 1920.
4. Geological Survey of Canada, *Annual Summary Report*, 1909.
5. Board on Topographical Surveys and Maps, Interior File 17884. A report written by Surveyor-General F. H. Peters, 15 April 1925.
6. Board on Topographical Surveys and Maps, Interior File 17884. A paper written by Deputy Surveyor-General T. Shanks, 24 March 1924.
7. Geological Survey of Canada, *Annual Summary Report*, 1925, Part A, p. 234.
8. Board on Topographical Surveys and Maps, Interior File 17884. Letter Collins to Peters, 8 May 1925.
9. *Ibid*. Letter Peters to Collins, 8 May 1925.
10. *Ibid*. Letter Shanks to Boyd, 21 March 1924.
11. *Ibid*. Memorandum Shanks to Deville, 23 April 1924.
12. Standing Joint Committee on Map Design and Standardization, Surveys and Mapping Branch, File 1135–M2–1. Minutes of meeting, 10 July 1951. The Military started publishing 1:50,000 maps in 1950, Topographical Survey waited until 1951.
13. Additional information on the symbolization used on the maps is given in: *Manual for Topographers*, Topographical Survey of Canada, Interior, 1927, 2nd edition 1936; *The Standard Mapping Method of the Geographical Section G. S. Canada*, National Defence, 1936.

13a. *Manual for Topographers*, Topographical Survey of Canada, Interior, 1927, p. 75.

14. This data is included in a report written by Surveyor-General Deville, Board on Topographical Surveys and Maps, Interior File 17884, 7 September 1920.
15. This account was taken from *Catalogue of Maps, Plans and Publications*, Legal Surveys and Map Service, Mines and Resources, 7th edition 1939.

Chapter 5 The 1:50,000 Series

1. Standing Joint Committee on Map Design and Standardization, Surveys and Mapping Branch File 1135–M2–1. Minutes of 10 July 1951. Details on the changes in sheet numbering caused by the change from one inch to 1:50,000 mapping is given in A. M. Floyd, *Historical Development of the National Topographic System of Map Numbering for Canada*, Surveys and Mapping Branch, Energy, Mines and Resources, 1969. Half sheets were extended to 62° north latitude in 1954 because of the change in the design of the map border of the 1953 standard.
2. *Catalogue of Published Maps*, Map Compilation and Reproduction Division, Surveys and Mapping Branch, Mines and Technical Surveys. Published annually from 1961 to 1970, then in 1972, 1974 and 1978. Since 1978 information has been on microfiche.
3. Standing Joint Committee on Map Design and Standardization, Surveys and Mapping Branch File 1135–M2–1. Minutes of 15 July 1953.

4. *Ibid*, Minutes of 16 June 1952.
5. Preliminary and provisional styles of mapping were not processed through the Committee on Map Design and Standardization and as a result the reasons for their style peculiarities are not documented.
6. Details on symbolization are given in : *Standard Conventional Signs and Colours*, Army Survey Establishment, National Defence, 1952, 2nd edition 1958; *Cartographic Manual*, Map Compilation and Reproduction Division, Surveys and Mapping Branch, Mines and Technical Surveys, 1954; *Interpretation and Classification Instructions*, Army Survey Establishment, R.C.E. National Defence, 1958; *1:25,000 Drafting Specifications*, Army Survey Establishment, National Defence, 1962; *1:50,000 Drafting Specifications*, Army Survey Establishment, National Defence, 1963; *1:250,000 Drafting Specifications*, Army Survey Establishment, National Defence, 1963; *Topographical Survey Provisional Manual of Compilation Specifications*, Surveys and Mapping Branch, Mines and Technical Surveys, 1965; *Topographical Survey Provisional Manual of Compilation Specifications*, Surveys and Mapping Branch, Mines and Technical Surveys, 2nd edition 1967, 3rd edition 1974.
7. The military authorities were the traditional supporters of the differentiation between deciduous and coniferous growth. The break with this tradition came in 1958 with the publication of *Interpretation and Classification Instructions*, Army Establishment, R.C.E. National Defence, 1958, which specifies (on p. 43) that a single symbol will be used for wooded areas. The Topographical Survey quickly followed suit.
8. See also *Instructions Générales d'arpentage*, Direction Générale du Domaine Territorial, Ministère des Terres et Forêts, Québec 1971.
9. Full details are given in: *Manual Relating to Surveys and Surveyors*, Association of Ontario Land Surveyors, Toronto (1st edition 1959, 2nd edition 1973); *Manual of Topographic Map Manuscript Editing*, Surveys and Mapping Branch, Mines and Technical Surveys, 2nd edition 1967 (3rd edition 1974), pp. 47–50.
10. *Survey Systems within the Crown Domain*, Surveys and Mapping Branch, Victoria 1975.
11. Topographical Survey of British Columbia, Interior File 18884, Letter from Surveyor-General of British Columbia dated 1 March 1937.
12. Overprint revision was first used, experimentally, by the Army Survey Establishment in 1963. Originally only new highways were thus shown, and the standard red ink was used. As this caused confusion as to what roads had been added by overprint, the colour was changed to purple in 1964. (See Standing Joint Committee on Map Design and Standardization, Surveys and Mapping Branch File 1135–M2–1, Minutes of 18 December 1964.)

Chapter 6 The Two-Mile and 1:125,000 Series

1. Major E. H. Hills, *Report on the Survey of Canada*, War Office, London 1903, pp. 15–16.
2. *Catalogue of Published Maps*, Map Compilation and Reproduction Division, Surveys and Mapping Branch, Mines and Technical Surveys, 1978, p. 52.
3. *Ibid*, p. 51.

Chapter 7 The Four-Mile Series and the 1:250,000 Series

1. For a complete description of the method see 'Photographic Surveying' in Bulletin No. 56, Topographical Survey, Interior, 1924.
2. See also 'The use of Aerial Photographs for Mapping' in Bulletin No. 62, Topographical Survey, Interior, 1932.
3. See also *Manual for Plotting Trimetrogon Photographs*, Hydrographic and Map Service, Mines and Resources, 1944.
4. For a description of this technique see 'The use of Aerial Photographs for Mapping' in Bulletin No. 62, Topographical Survey, Interior, 1932, pp. 50–61.
5. Standing Joint Committee on Map Design and Standardization, Surveys and Mapping Branch File 1135–M2–1. Minutes of meeting of 10 January 1950.
6. For a full description of 1:250,000 symbols see: *Interpretation and Classification Instructions*, Army Survey Establishment, R.C.E., National Defence, 1958; *1:250,000 Drafting Specifications*, Army Survey Establishment, National Defence, 1963; *Topographical Survey Provisional Manual of Compilation Specifications* Surveys and Mapping Branch, Mines and Technical Surveys, 1965; *Topographical Survey Provisional Manual of Compilation Specifications*, Surveys and Mapping Branch, Mines and Technical Surveys, 2nd edition 1967, 3rd edition 1974.

Chapter 8 The Eight-Mile Series and Canadian Aeronautical Charts

1. Board on Topographical Surveys and Maps, Interior File 17884. Memorandum written by Deville, 9 January 1923.
2. Aerial Surveys 1931 and 1932, Interior File 20704.
3. A complete description of this method is in: *Manual for Plotting Trimetrogen Photographs*, Hydrographic and Map Service, Mines and Resources, 1944.

Chapter 9 The 1:25,000 series

1. Board of Topographical Surveys and Maps, Interior File 17884. Report by Colonel J. Sutherland Brown, 3 April 1925.
2. Standing Joint Committee on Map Design and Standardization, Surveys and Mapping Branch File 1135–M2–1. Minutes of meeting 13 January 1953.
3. Complete details on symbol specifications are given in: *Standard Conventional Signs and Colours*, Army Survey Establishment, National Defence, 1952 (2nd edition 1958); *Cartographic Manual*, Map Compilation and Reproduction Division, Surveys and Mapping Branch, Mines and Technical Surveys, 1954; *Interpretation and Classification Instructions*, Army Survey Establishment, R.C.E., National Defence, 1958; *1:25,000 Drafting Specifications*, Army Survey Establishment, National Defence, 1962; *Topographical Survey Provisional Manual of Compilation Specifications*, Surveys and Mapping Branch, Mines and Technical Surveys, 1965, 2nd edition 1967, 3rd edition 1974.
4. Committee on Map Standardization, Surveys and Mapping Branch File 1135–M2–1. Minutes of meeting, 15 July 1964.

Chapter 10 Maps of the Chief Geographer's Office

1. Sheets of the 1:250,000 and 1:500,000 series are described and a list of published sheets given in: *Catalogue of the Maps in the Collection of the Geographic Board*, The Geographic Board of Canada, 1922; *Catalogue of Maps, Plans and Publications*, Legal Surveys and Map Service, Mines and Resources, 7th edition 1939.
2. Copies of this map were bound into the annual report of the Topographic Survey for 1928.
3. Complete details on the specification for IMW maps may be found in *Specifications of the International Map of the World on the Millionth Scale (IMW)*, United Nations, New York 1962.

Chapter 11 Federal Thematic Maps

1. The addresses of the map-producing agencies mentioned in this chapter and the catalogues they issue are listed in Appendix 8.
2. Although this chapter is devoted to federal thematic mapping, for convenience federal, provincial and territorial road maps are described here.
3. Again for convenience, federal, provincial and territorial electoral maps are described in this section.

Chapter 12 Hydrographic Charts

1. C. E. Heidenreich, 'Explorations and Mapping of Samuel de Champlain, 1603–1632', *Cartographica* 17, 1976, p. 79.
2. *Ibid*, p. 88.
3. Don W. Thomson, *Men and Meridians: The History of Surveying and Mapping in Canada*, Vol 1, Queen's Printer, Ottawa 1967, p. 108.
4. Ruth McKenzie, 'Admiral Bayfield, Pioneer Nautical Surveyor', Information Canada, Ottawa 1976.
5. N. L. Nicholson, 'The Boundaries of the Canadian Confederation', Macmillan, Toronto 1979.
6. Don W. Thomson, *Men and Meridians: The History of Surveying and Mapping Canada*, Vol 2, Queen's Printer, Ottawa 1968, pp. 209.
7. R. W. Sandilands, 'The History of Hydrographic Surveying in British Columbia', *The (Canadian) Cartographer*, 7, 1970, pp. 126–130.
8. *Symbols and Abbreviations used on Nautical Charts*, Fisheries and the Environment, Canadian Hydrographic Service, Ottawa 1978.
9. Colin H. Martin, 'Navigational Charts for Pleasure Boating', *The (Canadian) Cartographer*, 6, 1969, pp. 46–53.

Chapter 13 Important Provincial Map Series

1. The addresses of the map-producing agencies mentioned in this chapter and the catalogues, brochures, etc., that they issue are listed in Appendix 8.

Chapter 14 Atlases

1. *Electoral Atlas of the Dominion of Canada as Divided for the Revision of the Voters' Lists made in the Year 1894*, Government Printing Bureau, Ottawa 1895. For this and other federal electoral maps see also Joan Winearls, 'Federal Electoral Maps of Canada, 1867–1970', *The (Canadian) Cartographer*, 9, 1972, pp. 1–24.
2. N. L. Nicholson, 'A Survey of Single Country Atlases', *Geographical Bulletin*, 2, 1952, pp. 19–35.
3. N. L. Nicholson, 'Canada and the International Geographical Union', *The Canadian Geographer*, 14, 1959, pp. 37–41.
4. J. White (ed.), *Atlas of Canada*, Department of the Interior, Ottawa 1906.
5. J. E. Chalifour (ed.), *Atlas of Canada*, Department of the Interior, Ottawa 1915.
6. B. Brouillette, *Atlas of Canada Project*, Social Science Research Council of Canada, Ottawa 1945.
7. N. L. Nicholson (ed.), *Atlas of Canada*, Department of Mines and Technical Surveys, Ottawa 1958.
8. N. L. Nicholson, 'Some Elements in the Development of the National Atlas of Canada', *Geographical Bulletin*, 16, 1961, pp. 45–61.
9. G. Fremlin (ed.), *The National Atlas of Canada*, Macmillan, Toronto 1974.
10. N. L. Nicholson, 'Canada in Six Atlases', *The (Canadian) Cartographer*, 7, 1970, pp. 126–130; N. L. Nicholson and J. Fremlin, 'An Analysis of Canada's Provincial Atlases' in *Proceedings of the Plenary Session of the Commission on National Atlases*, Geodetic and Cartographic Society, Budapest 1962, pp. 141–169.

Chapter 15 Projections and Reference Systems

1. Interior 16555 Map Projections. Memorandum from H. Parry, Chief Cartographer, to Surveyor-General F. H. Peters, 28 September 1925.
2. National Defence HQC 5768, Correspondence with Department of the Interior, Vol. 2, Memorandum from Deville to Lt. Col. H. W. Brown, 10 November 1922.
3. *Ibid.* Letter from Major E. L. M. Burns (Officer Commanding the Geographical Section, General Staff) to Noel Ogilvie, Director of the Geodetic Survey, 27 February 1934.
4. J. A. Steers, *An Introduction to the Study of Map Projections*, 15th edition, University of London Press, London 1970, p. 189.
5. This study was started in 1955—see Committee on Map Standardization, Surveys and Mapping Branch File 1135-M2-1. Letter from T. H. Kihl (member of the Committee) to the Chairman (D. R. Pallett), 13 October 1955. The matter was studied for 16 months and was resolved at the meeting of 26 April 1957.

Chapter 16 Map Printing Methods and Map Accuracies

1. Wilhelm Weber, *A History of Lithography*, McGraw-Hill, Toronto 1965, p. 70.
2. *Manual for Topographers*, Topographical Survey of Canada, Interior, 1927, (2nd edition 1936), p. 13.
3. 'Investigation of Positions of Monuments on Base Lines and Meridians', Interior, Bulletin No. 57, 1921, p. 11.
4. This date is inferred. *Manual of Instructions for the Survey of Dominion Lands*, Interior, 8th edition 1913, gives full instructions for contract surveys on pages 131–133. The 9th edition, dated 1918 but signed by the Surveyor-General in April 1917, gives no instructions for contract work.
5. L. M. Sebert, *Mapping Procedures for Canadian One-Inch Topographical Maps Prior to World War II*, Surveys and Mapping Branch, Department of Energy, Mines and Resources, Ottawa 1970, p. A1.
6. This fact is stated in the margin of topographic sheets published in 1906.
7. Personal communication between L. M. Sebert and Mr Donald MacKay, a surveyor who took part in this survey.
8. For full details on block adjustments see J. A. R. Blais and M. R. Gareau, *'Program SPACE-M (for Photogrammetric Block Adjustment)': User's Guide*, Surveys and Mapping Branch, Department of Energy, Mines and Resources, Ottawa 1977; J. R. Blais, *Program SPACE-M: Theory and Development*, Surveys and Mapping Branch, Department of Energy, Mines and Resources, Ottawa 1977.
9. *A Guide to the Accuracy of Maps*, Surveys and Mapping Branch, Department of Energy, Mines and Resources, Ottawa 1976, pp. 4–6.

10. *Topographical Survey Map Evaluation Report* (information current as of June 1978), Surveys and Mapping Branch, Department of Energy, Mines and Resources, Ottawa 1978.
11. The Doppler Effect (named after the Austrian physicist Christian Johann Doppler, 1803–1853) describes the change in frequency of sound, radio or light waves observed when the emitter of the waves moves toward or away from the observer. Most people have observed the Doppler shift at some time or another without realizing the cause. For example, a person standing beside a railway track will hear the tone of the engine whistle drop as the train passes at high speed. The sound waves sent out by the whistle radiate at a constant frequency, but if the emitter of the sound rushes toward the observer, the waves reach the observer's ear at a higher frequency than that of emission. In contrast, as the sound emitter passes the observer and moves away, the waves reach the observer's ear at a lower frequency, and the sound registers at a lower pitch. Navigation satellites emit radio waves which exhibit the same shift on recording instruments at ground stations as the satellite passes overhead. By knowing the exact orbit of the satellite, and by measuring the change in signal frequency as the satellite passes, an exact distance can be computed from the ground station to the satellite at a number of positions in its orbit. By using a number of such recordings, observed over several days, the position of the ground station can be computed to an accuracy of less than a metre. A more technical description of this phenomenon can be found in: T. P. Gill, *The Doppler Effect*, Logos Press, London 1965.

Chapter 17 Geographical Names on Canadian Maps

1. See *Gazetteer of Nova Scotia*, 1978, p. xx.

A Short Glossary of Mapping Terms

Air Profile Recorder — An electronic device mounted in an aeroplane which provides a continuous profile of the terrain below. It is used to provide basic data for contouring.

Cased Road — A map symbol for a road consisting of two parallel black lines (the casing) with a colourful infill.

Control — The collective term used for the pattern of surveyed points that have been placed in the area being mapped to 'control' the position and scale of the map detail. Horizontal control is the array of points of known latitude and longitude; vertical control is the array of points of known elevation.

Multiplex — This was the first of the photogrammetric plotters to be used in Canada. In its basic form it consists of two optical projectors which throw images from slides of overlapping air photos down onto a drafting table. By using a red filter in one projector and a blue-green filter in the other a three dimensional view of the common overlap is obtained.

Order of Surveys — First order surveys are those of high precision, and as a consequence are expensive. Second, third and fourth order surveys are successively less precise and less costly.

Photomap — A map on which a mosaic of air photos replaces the usual map linework. An orthophoto map is a photomap on which the air photographs have been carefully rectified to remove the scale distortion often found in photographs particularly those of hilly or mountainous terrain.

Photogrammetry — The science of drawing maps from air photos.

Planimetric Map — A map without contours or other indication of relief.

Tellurometer — An electronic device for measuring distances. Measurements are obtained by timing the passage of electromagnetic waves over the lines being measured. Accuracies within a few centimetres are obtained for lines up to 30 km in length.

Traversing — A method of surveying in which the angles and distances along a given course are measured. In stadia traversing the distances are measured optically by theodolite to provide fourth order control.

Triangulation — A method of surveying in which signals (targets) are set out on hills or towers in the form of triangles. The angles, and in recent years the sides as well, of these triangles are measured to provide accurate positions for the signals.

Trilateration — The same as triangulation but only the lengths of the sides of triangles are measured.

For terms not included in this glossary see the references given in the index.

Index

BIBL. UNIV. LAUR. UNIV. LIB.
3 0007 00260054 1

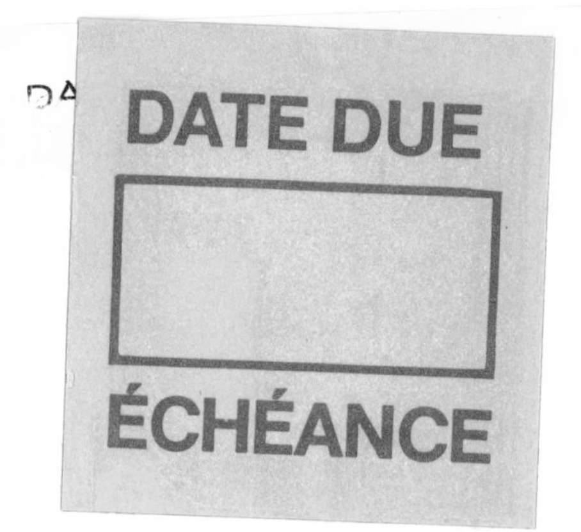
DATE DUE
ÉCHÉANCE